WHAT'S MINE IS MINE, WHAT'S YOURS IS MINE

THE USFSPA

AN ILLEGAL LAW ~ A CRIME AGAINST VETERANS

MSgt Andrew A. Bufalo USMC (Ret)

Copyright © 2010 S&B Publishing

This book may not be reproduced in whole or in part without the written permission of the author or the publisher.

ISBN 978-0-9845957-2-3

First Printing – November 2010
Printed in the United States of America

www.AllAmericanBooks.com

What's Mine is Mine, What's Yours is Mine

The shortest sentence in the English language is "I AM."

The *longest* is…"I DO."

What's Mine is Mine, What's Yours is Mine

This book is dedicated to all of the Patriots who have unselfishly served our great Nation, and asked for so little in return.

What's Mine is Mine, What's Yours is Mine

TABLE OF CONTENTS

For more information and the most recent developments visit the ULSG website at:

www.ULSG.org

DIVORCE IN GENERAL
Not the Same as a General Getting Divorced

Married couples are getting divorced in this country at an alarming rate, with more than fifty percent of all marriages ending in divorce. Armed Forces marriages are particularly vulnerable, due to the unique nature of military service.

No one gets married contemplating the possibility of divorce, but in this day and age it is just a fact of life. Building a strong marriage is difficult, and that is especially true in the armed forces. Long deployments, frequent moves, combat tours and a regimented military lifestyle all place a tremendous strain on most military families. Sometimes it is the military member who submits to the pressure, but the trend in recent years has been for the spouse to take the "easy way out" when confronted by these challenges. They often feel abandoned while the military member is deployed, fear for their soldier's safety during periods of combat, grow tired of setting up new households every few years, and are sometimes tempted to engage in extramarital affairs while their military spouses are away. With all that, it's amazing that *any* marriages survive.

In recent years the divorce option has become much more attractive for the spouses of career military members due to a law known as the Uniformed Services Former Spouses Protection Act, because they know they don't have to leave empty handed – regardless of who is at fault for the failure of the marriage.

A number of experts have opined that marriage and military service are not compatible, and to some degree they are right. In other eras soldiering was a comparatively monastic profession, with low pay and Spartan living conditions (and *all* of the troops were male). That remained true throughout much of this century, and has only begun to change in recent years. As a result, today's soldiers are expected to be the expeditionary warriors they have always been, while *somehow* cultivating a "normal" family life at the same time. More times than not the women (and some men) who marry into this lifestyle have no idea what they are getting themselves into – and when they decide to "abandon ship" the military member often ends up paying for their spouse's mistake because of the life preserver provided by USFSPA.

THE PROBLEM
The Pendulum Has Swung Too Far

The USFSPA was written for a different era - when women rarely worked outside the home, dual income families were unusual, and there were few opportunities for military spouses to pursue an education and become independent. Times have changed, and so should the USFSPA.

Most people would agree that both members of a military marriage which ends in divorce should be treated fairly and equitably in the courts. Even in this era of no-fault divorces, there is no legal prohibition against basing divorce settlements on merit, need, and ability to pay. Unfortunately, a review of court experience with the USFSPA since 1982 shows that it has operated in theory as an *option*, but has operated in practice as a *mandate*.

An overwhelming majority of the national community of military veterans' organizations, with an aggregate membership of more than 10,000,000, support reform of the USFSPA to resolve inequities. Some (e.g., the American Legion) even advocate its repeal. After twenty years, the

11

egregious impact of the USFSPA is in plain view. Moreover, the status of the military spouse of the early 1980s has undergone revolutionary social, economic, and cultural changes, which have moderated the climate in which the USFSPA was conceived.

Honorable military veterans have never subscribed to the theory of the "throwaway wife" as espoused by feminist-interest groups during the time the USFSPA was being considered by Congress. Neither do they conform to the stereotypical military member then represented as a quintessential, devil-may-care, madcap playboy who roamed the earth seeking adulterous relationships while leaving at home a faithful wife and children who can be abandoned at will.

Under the USFSPA, former spouses are not given an automatic right to receive an allocation of retired pay. Rather, retired pay *can*, at the discretion of the State courts, be treated either as the sole property of the member or as the property of the member and his or her spouse – or at least that was how it was *supposed* to work.

THE McCARTY DECISION
The Men in Black got this one right...

On June 26, 1981 the U. S. Supreme Court ruled in the *McCarty v. McCarty* decision that military retired pay is **not** a vested property right and therefore was **not subject to division in a divorce**, and went on to say that "the military retirement system **confers no entitlement to retired pay upon the retired member's spouse**, and does not embody even a limited 'community property' concept." That sounds pretty clear, so why then does military retainer pay continue to be divided?

Supreme Court Justice Harry A. BLACKMUN wisely wrote the following when he delivered the majority 6-3 opinion of the Court on *McCarty v. McCarty*:

"The fundamental purpose of retired pay of military personnel is to provide for the national defense. It is this fundamental purpose which requires the finding that there is a federal interest to protect in the case before the Court, that the federal interests dictates a finding that *retired pay is not a vested property right.* The findings by scattered state courts

13

that retired pay is a vested property right, based upon certain characteristics of the payment, pose significant threats to the special nature of this 'entitlement' and its function in national defense."

"(a) There is a conflict between the terms of the federal military retirement statutes and the community property right asserted by appellee. *The military retirement system confers no entitlement to retired pay upon the retired member's spouse, and does not embody even a limited 'community property concept.* Rather, the language, structure, and history of the statutes make it clear that retired pay continues to be the personal entitlement of the retiree."

"(b) Moreover, the application of community property principles to military retired pay threatens grave harm to "clear and substantial" [453 U.S. 210, 211] federal interests. Thus, the community property division of retired pay, by reducing the amounts that Congress has determined are necessary for the retired member, has the potential to frustrate the congressional objective of providing for the retired service member. In addition, such a division has the potential to interfere with the congressional goals of having the military retirement system serve as an inducement for enlistment and re-enlistment and as an encouragement to orderly promotion and a youthful military."

That reasoning made a lot of sense then, and it makes even *more* sense today!

STEALTH AND DECEPTION
Taking the "Supreme" Out of the Supreme Court

The **USFSPA** was enacted without debate, and was passed almost unnoticed as a component of a larger Defense Appropriations bill. It was also <u>backdated to the day before the *McCarty* ruling,</u> thereby circumventing the Supreme Court. The bill's proponents claim the law "recognizes that *both spouses contribute <u>equally</u>* to the service member's ability to earn a wage and receive a pension." That is simply ridiculous.

Congress acted shortly after the *McCarty* decision by enacting the Uniformed Services Former Spouses' Protection Act. The USFSPA, or FSPA or FSVA, is found at 10 U.S.C. §§1408 et seq. (1982), effective date February 1, 1983, retroactive to June 25, 1981, one day prior to the *McCarty* decision. Sponsored by Rep. Patricia Schroeder (D-CO), the FSPA in effect reversed the McCarty decision, rejecting the Court's concerns regarding military retention, enlistment, and the economic needs of older veterans. Also known as the Former Spouse Victim Act by military retirees, the FSPA has been a source of confusion and controversy at both the

15

state and national level.

The FSPA applies to the "uniformed services," defined to include the Army, Navy, Air Force, Marine Corps, Coast Guard, commissioned corps of the National Oceanic and Atmospheric Administration, and the commissioned corps of the Public Health Service. The FSPA applies to active duty, retired, and reserve/guard (whether active duty, inactive status, or retired), in pay and non-pay categories.

Since the FSPA is a federal statute, its provisions and the regulations thereunder preempt or supersede state laws. A state court order that contradicts the FSPA will not be enforceable. The FSPA, with limitations, allows state courts to treat a military pension either as property solely of the service member, or as property of the member and his or her spouse in accordance with the law of the jurisdiction for pay periods beginning after June 25, 1981. In the unlikely event that a state court order divided military retirement pay before June 26, 1981, in conformity with the FSPA, that order will be honored.

The biggest misconception about USFSPA is that it *requires* the division of military retainer pay in the event of divorce, when in reality it simply *allows* it. There is a big difference between the words "may" and "must," but many family court judges miss that nuance when dividing property.

The thing which probably sticks in the craw of military retirees the most is the way USFSPA was backdated in order to sidestep a clear ruling by the highest court in the land. To put it in perspective - *just imagine what would happen if Congress were to pass a law reinstating segregation, and backdated it to one day before 'Brown v. Board of Education!'*

THE RULE OF PROPERTY
You Can't Take It With You...Can You?

Since military retirees have no property rights to military "retired" pay, then why does a former spouse gain a property asset right?

The U.S. Supreme Court ruled in *Buchanan v. Alexander* (45 U.S.20 1846) that money owed by the United States to the individual service member belongs to the Treasury until it is paid to that individual. Essentially, the Supreme Court held that courts cannot tell a federal disbursing official what to do since it would defeat the purpose for which Congress appropriated the money. If the specific reason Congress appropriated funds for the retired military member after twenty years of active duty is not as compensation for continuing military obligations, what then is the reason? What law provides other reasons? The Former Spouse Act provides that a former spouse may receive military retirement pay directly from a military finance center, without sending the money to the military member first.

In *United States vs Tyler* (105 U.S. 244, 1882), the Court characterized such pay as "compensation... which continued at a reduced rate." If it is indeed current compensation rather than property (as the Court ruled), its division is effectively "alimony for life."

So where did the words *deferred income* and *pension* come from?

On May 19, 1986 the Department of the Air Force Accounting and Finance Center notified a retired Air Force Officer that he owned the Federal Government $3,161.63. The letter stated that he was subject to the Dual Compensation Law, codified in title 5, United States Code (U.S.C.) section 5532. This retired military officer was required by federal law to pay back $3,161.63 of his military "retired" pay for teaching math to Navajo Indians. The job was considered federal employment, since his contract was with the Bureau of Indian Affairs, Department of the Interior.

Would a former spouse who receives military retired/retainer pay as a community property asset be required to return funds for teaching Navajo Indians math? The fact that this officer had to return this compensation proved that it was *not* a pension!

Another example occurred when a retired Marine had his "retired pay" halted while working in Saudi Arabia. Major Stephen H. Hartnett (USMC Ret) accepted a job in May of 1985 with the Delaware-based firm of Frank E. Basil Inc, and shortly thereafter had his military "retired" pay stopped because the U.S. Comptroller General ruled he was under the supervision and control of the Saudi Government. Long standing legal rulings have placed retired military personnel in this category, since they are subject to recall to active duty. (Source: *Air Force Times* Dated Sept 1, 1986)

Former spouses of military members do not lose their property rights to military retired pay if they become employed by a foreign government.

SFC Charles J. O'Fearna retired from the Army in 1965,

after 22 years of military service. He moved to Australia in 1966, and has lived there ever since. In April of 1981 O'Fearna became a naturalized citizen of Australia, apparently without realizing what effect it would have on his military "retired" pay. The Comptroller General held in previous decisions that military retirees lose their right to retired pay when they lose their citizenship. The Comptroller General stated that all members on the retired list of the regular Army remain a part of that force, and are relied upon as a dependable source of manpower. The finance center told O'Fearna that his retired pay ended when he forfeited his citizenship. They stopped it immediately and advised him to refund retired pay received since April, 1981. (Source: *Air Force Times* dated March 5, 1984)

Not only do former spouses not lose their so-called property asset rights if they lose American Citizenship – many of them are not and *never were* U.S. citizens! Millions of dollars of military retired/retainer pay is being paid to Foreign Nationals who divorce military members, and many of them have even returned to their home countries!

***Costello vs. United States*, Constitutional law 278.6 (1):**

"Military retirement pay is not deferred compensation for past services but, like active duty pay, is pay for continuing military service and as such, can be prospectively altered without offending due process."

It is pretty clear that military retainer pay is *not* the property of military retirees, because if it *was* it couldn't be taken away by the government. It is also clear that the same rules do not apply to the former spouses. Why are we sending checks to foreign nationals who live in other countries, while depriving the former service members who served our country?

THE RETIREMENT MYTH
Uhh...We're Not Done With You Just Yet!

Military retirees do not have 'property rights' to their retired pay. It is earned based upon on a day-by-day availability to serve if recalled. So how can something they don't own be divided?

It is a little known fact that service members do not actually "retire," but are instead TRANSFERRED to the inactive reserve! For that matter, military "retirees" *can be and have been recalled* to active duty, and remain subject to recall for life (up to a specified age).

In *Lemly vs. United States* (1948) it was ruled that "retirement pay is a continuation of active pay on a reduced basis. Even though an officer is retired from active duty and is receiving retirement pay, he is still subject to recall to active duty as long as his physical condition will permit. He is still an officer in the service of his country even though on the retired list."

It is not only retirees who can be called back to duty. Underscoring how the wars in Iraq and Afghanistan have stretched the military's resources, call-ups included 5,674 members of the Army's Individual Ready Reserve, or IRR. Those are soldiers who were honorably discharged after finishing their active-duty tours - usually at least four to six years - but remain part of the IRR for the rest of the eight-year commitment they make when they join the Army.

The IRR call-up is the first major one in the thirteen years since 20,277 troops were ordered back to duty for the Persian Gulf War, and many of those who were brought back were retirees with critical skill sets.

Military retirement regulations governing recall to active duty are as follows:

In an event where Congress declares a state of war or national emergency, the Secretary of Defense can authorize the Secretaries of the Army, Navy, and Air Force to recall retired military personnel. Retirees may be recalled up to age 64 for general officers, age 62 for Warrant Officers, and age 60 for all others. Retirees are placed into one of three categories for recall purposes.

➢ Category I (usually called first) includes retired service members who meet the age and grade criteria, were not retired for permanent disability, have a U.S. address, and have been retired fewer than five years.

➢ Category II (usually called after Category I) includes service members with the same qualifications as the first category who have been retired for more than five years.

➢ Category III includes all other retired service members, including permanently disqualified disability retired service members. Category III will generally not be recalled.

A member of the Retired Reserve who returns to active duty may only receive one type of payment, therefore once their normal pay is reinstated, their "retired pay" stops. So how can it be considered property?

According to Department of Defense (DOD) Directive 1352.1:

Involuntary Orders to Active Duty. The Secretary of a Military Department may order any retired Regular member, retired Reserve member who has completed at least 20 years of active military service, or a member of the Fleet Reserve or Fleet Marine Corps Reserve *to active duty without the member's consent at any time to perform duties deemed necessary in the interests of national defense* in accordance with 10 U.S.C. 683 (reference (b)). This includes the authority to order a retired member who is subject to the Uniform Code of Military Justice (UCMJ) to active duty to facilitate the exercise of court-martial jurisdiction under Section 302(a) of reference (b). A retired member may not be involuntarily ordered to active duty solely for obtaining court-martial jurisdiction over the member.

Proponents of USFSPA often cite the need for "equity" and "fairness." Can spouses be involuntarily ordered to active duty? Of course not! So what is *equitable* and *fair* about that?

THE UCMJ
How About a Uniform Code of <u>Spousal</u> Justice?

Military retirees are subject to the applicable provisions of the Uniform Code of Military Justice (UCMJ), and can lose their retirement benefits for violating it. Spouses (and ex-spouses) are not accountable, and *never were.*

If a military retiree commits a crime and spends more than six months in a correctional facility, that military retiree will lose his or her military "retirement" for the duration of time spent in jail. Once released, the retiree must reapply to DFAS and request reinstatement of their regular retirement income.

Take for example the case of Lieutenant Colonel Oliver North, who lost his Military retired/retainer pay for being convicted of a crime. It took an act of Congress to restore his retired pay. If Congress had not acted, then are we to assume that Colonel North's wife would have *also* lost her property rights to his pay in the event of a divorce?

The answer is NO, because this federal law does NOT apply to a former spouse who is collecting up to half of that retiree's monthly military retirement check as part of a community property asset. In fact, in the event the spouse is the one convicted of a crime, he or she can legally continue to receive their court awarded portion of a military "retirement" check, no matter how long he/she spends behind bars.

I'm sure you're asking, "Why is that the case?" The answer is simple. Members of the Armed Forces continue to be subject to provisions of the Uniform Code of Military Justice even after retirement. Why? Because they are actually in a reserve status, and subject to recall to active duty – while their former spouses *never were* answerable to the UCMJ and *cannot be recalled* since they *never served*!

You may remember the infamous Kelly Flynn case of a few years back. Lt. Flynn was the Air Force's first female B-52 pilot, but unfortunately she was also an unmarried officer who was having an affair with a married civilian. Lt. Flynn was advised by a First Sergeant, and later ordered by her Commander, to terminate the affair. She broke up with her "boyfriend," but later they got back together, and when asked about it Lt. Flynn lied. She was then charged with the offenses of adultery, giving a false official statement, conduct unbecoming an officer, and disobeying an order of a superior commissioned officer.

So, where was the "military connection" for the adultery charge? Well, the civilian "boyfriend" was the husband of an active duty enlisted Air Force member stationed at the same base as Lt. Flynn. Therefore, Lt. Flynn's "affair" had a direct negative impact on the morale of that military service member (the enlisted wife is the one who originally complained about the inappropriate actions of Lt. Flynn).

Lt. Flynn didn't face a military court however, but was instead allowed to resign her commission in lieu of court martial (lots of media attention probably had something to do with this decision by the Air Force). Of course, the married man in the middle of all this *couldn't* be disciplined, because since he was only a *spouse* he was not subject to the UCMJ.

Another egregious example of how the UCMJ applies only to military members follows, and it is quite ironic. A female sailor married a male enlisted shipmate who brought two children from an earlier marriage into the relationship, and shortly thereafter her husband was discharged from the Navy for drug abuse. Subsequently they had a child of their own, with the wife being the principal family provider throughout most of the marriage. They were divorced in 1994, and when she retired as a First Class Petty Officer with twenty years of service her husband, who had remarried and had been in and out of jail, was awarded 30% of her retired pay for life. The irony, of course, lies in the fact that the husband was prosecuted under the UCMJ and kicked out of the Navy, and would as a result never be entitled to a military retirement of his own – and then he ended up with a share of *hers!*

TAKING THE OATH
Or, Saying "I Do"

Men and women enlisting in the Armed Forces must take an oath, and must do so again each time they re-enlist over the course of their careers, in order to become eligible for retirement. Their spouses only have to say "I do" - *once* - in order to qualify for the very same benefits.

Each time a service member enlists or re-enlists in the Armed Forces they are required to raise their right hand and swear the following oath:

"I do solemnly swear (or affirm) that I will support and defend the Constitution of the United States against all enemies, foreign and domestic; that I will bear true faith and allegiance to the same; and that I will obey the orders of the President of the United States and the orders of the officers appointed over me, according to regulations and the Uniform Code of Military Justice. So help me God."

Spouses have no such obligation, and can opt-out of their vow of "I do" at any time – while the service member must remain in the Armed Forces for the duration of their contracts and must complete a minimum of twenty years in order to retire. This is "equal?"

BASIC TRAINING IS NO HONEYMOON
Parris Island is <u>Not</u> Fantasy Island

"Listen up, maggot!"

Military recruits must endure the rigors of basic training as the first step toward retirement, while their spouses obviously do not. That may not sound like a big deal to those who have never been to Parris Island (or Great Lakes, Fort Jackson, etc.), but I assure you that it *is* when you consider the former spouse's argument that they "make an equal contribution" toward the military member's successful career.

Comparing what a service member must go through in order commence a military career to what a spouse must do to enter into his or her "career" is like comparing the movie *Full Metal Jacket* to *Honeymoon in Vegas*. There is no comparison! So tell me again how military spouses "make an equal contribution?"

Basic Training is only the beginning. The following chapters will take an in-depth look at the concept of "equal contribution" throughout a military career.

27

EQUAL CONTRIBUTION I
In the Field vs. At the Mall

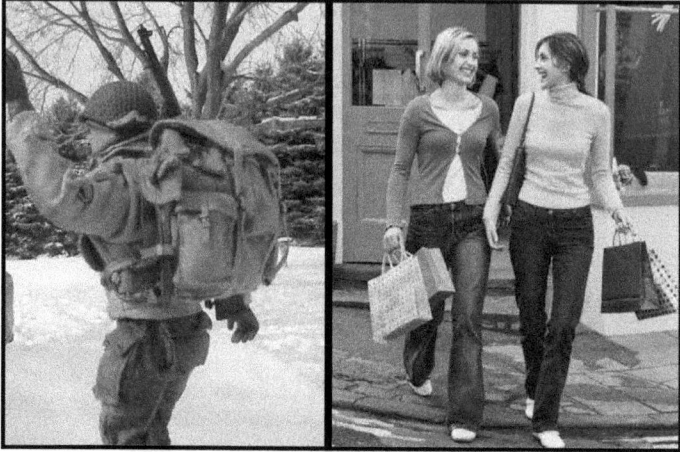

**Carrying a pack in the field should not be confused
with carrying shopping bags at the mall!**

One of the things which really sets military service apart
from the private sector is the need to participate in "field
exercises." In the course of my military career I participated
in everything from a Combined Arms Exercise in the Mojave
Desert (where the temperature exceeded 140 degrees), to a
cold weather operation above the Arctic Circle in the dead of
winter, and everything in between. As I recall the packs were
always heavy, the marches long, and the ground upon which
I slept cold and hard.

In every instance my then-wife made her "contribution"
by dropping me off at the battalion area before turning
around and heading to the mall for a day of shopping.

Equal contribution?

EQUAL CONTRIBUTION II
Deployments, And Those We Leave Behind

Military spouses often bring up the long stretches of time they are left behind while military members go on deployments, but what about those of us who actually had to GO? During the year I was stationed in Okinawa, for example, my wife lived her life exactly as she had before – while I lived in a barracks room with another sergeant and counted the days until I returned to the "world" on the "freedom bird." Also keep in mind the vast majority of my paycheck was being sent home to provide for my spouse during that entire time.

Equal contribution?

EQUAL CONTRIBUTION III
Three Hots and a Cot

Which one do you think the *spouse* sleeps in?

Throughout the nine-plus years of my marriage, my wife never spent *one single night* without a roof over her head and a comfortable bed to sleep in. Not one. In the meantime I slept in tents, in "snow graves, on armored vehicles, and crammed into troop compartments aboard Navy ships where the "racks" were stacked four high. When she became unhappy with base housing, I even went out and bought a new house so she would be comfortable while I was deployed (and stop complaining).

Equal contribution?

Once again, where does the *spouse* sleep?

What's Mine is Mine, What's Yours is Mine

Show me the spouse who has slept under conditions such as these, and I'll show you one who *might* be entitled to a *portion* of someone's retainer pay!

EQUAL CONTRIBUTION IV
WIAs and KIAs in the Commissary

How many spouses have returned home in a flag draped coffin?

As I write this, there have been over four thousand members of the United States Armed Forces killed in action in Iraq. During the Vietnam War, over fifty thousand of our troops fell. In World War Two, the number was in the vicinity of half a million. Throw in Korea, the Civil War, and all of the other wars and conflicts we have participated in throughout the course of our history, and the number becomes quite staggering. In the meantime some spouses faithfully waited at home, some waited without being faithful, and yet others walked out altogether - but none of them died in combat.

No spouse has ever made the supreme sacrifice for this nation, and yet they somehow they feel entitled to a share of what is earned by soldiers during a lifetime spent placing themselves in harm's way.

Equal contribution *indeed*!

32

FORMULAS FOR DISASTER
One of These States is Not Like the Other

The lack of clarity in the Uniformed Services Former Spouses' Protection Act (USFSPA) has caused a great deal of misinterpretation by the state courts. As a result, this ill-defined law places military members and their former spouses in a situation where there is no consistency among military divorces.

There is no consistency from a state-by-state/court-to-court standpoint, or from divorce to divorce. The divorces of fifty military members, with the exact same marital conditions, but whom are each stationed in a different state, will culminate in fifty different divorce scenarios. This treatment of military members, by reason of their transient employment, is discrimination.

As you will see in a later chapter, the root of the problem lies in the federal government deferring to the states, and in the states using a law passed by that very same federal government to justify *their* actions.

THE IRS
Emasculating the 550 Pound Gorilla

The I.R.S. Code 26 C.F.R. S 31. 3401 (a)-1(b) (1) (ii) states that military retired pay is a <u>Current Wage</u>, and it is taxed as such. If that is the case, (and the IRS is presumably always correct in such matters) how can it be divided as property during a divorce proceeding?

How can a state court classify military retired/retainer pay as a *deferred* income? Current wages are not a property asset, therefore the courts seem to have *changed* the classification of military retired/retainer pay in order for the pay to appear as a property asset.

For many years, State Courts did not require Former Spouses to pay taxes each month on the retired/retainer pay that they received as a community property asset - therefore the military member was being forced to pay federal taxes on income he or she never received. The IRS noted this problem and has since forced the Military Finance centers to send the former spouses a tax statement. *That took 20 years to get accomplished.*

34

The Internal Revenue Service (IRS), through the Federal Government's issue of W-2P Forms, identified military retired pay as separate from 'pensions and annuities.' And the Subchapter on Forfeiture of Annuities and Retired Pay, USC 5, 8311, defined military retired pay separately from federal employees' annuities. Even the *Encyclopedia Britannica* of the time noted that military retired pay was different from 'public employee pensions,' "... they (military retired pay measures) continued a certain portion of pay."

Once again, how can military retainer pay be classified as a current wage for the purpose of taxation, and as property in the divorce court? It seems like a case of having your cake and eating it too!

THE 20 YEAR OBLIGATION
For Some…

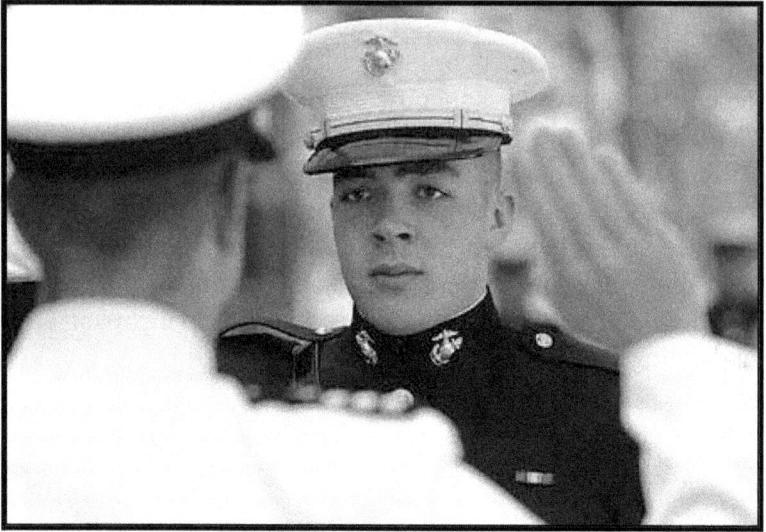

Members of the Armed Forces must complete a *minimum* of twenty years service in order to qualify for a "pension." Former spouses have no such requirement. Why the inequity?

Enlisted service members, including combat veterans, can be honorably - but involuntarily - discharged, and commissioned officers may resign, with as many as 19 years, 11 months, and 29 days service without any compensation or benefits forthcoming. They have served longer than ten years (as in the former spouses' proposals) yet are entitled to nothing from the Armed Forces for their sacrifices and hardships endured while serving their Nation.

Why, then, should spouses of less than twenty years be considered above the service rendered by these men and women, many of them combat veterans?

Military "retirement" pay is unlike civilian retirement pay systems. First and foremost, there is no "vesting" in the military retirement system. There are no special retirement accounts, no matching funds provision, no interest. You either qualify for retirement by honorably serving over twenty years in the military, or you do not. If you are discharged from the military with 19 years, 11 months, and 29 days of service, for example, you do not qualify for retirement pay (other than a few "early retirement" programs, which were designed to reduce the size of the armed forces).

On October 6, 1945 Public Law 190 - the Armed Forces Voluntary Recruitment Act of 1945 (H.R. 3951) - was passed to stimulate volunteer enlistments in the Regular Military and Naval Establishments of the United States. Section 4 of this law states, "Whenever any enlisted man of the regulator Army shall have completed not less than twenty or more than twenty-nine years of active services, he may upon his own request, be transferred and retired shall receive, except with respect to periods of active duty he may be required to performed, until his death, annual pay." There is like status for officer personnel. This law clearly states that the military member remains in a reserve status, and that the military member will remain so until death. Thus the reason for retainer pay - NOT a pension. It should also be noted that no such laws exist for the spouse.

So, the bottom line is a member of the Armed Forces must stay in the military for a minimum of twenty years in order to qualify for a "retirement." Why is there no such requirement for spouses, given their "equal contribution" argument?

WINDFALL PROFITS
A Reward For Time Not Served!

Payments to former spouses are based upon the military member's rank and years of service *at the time of retirement*, rather than *the date of divorce* as they should be. Why are former spouses being granted "windfalls" based upon a military member's continued service and promotions *after* the divorce?

The USFSPA does not require courts to base an allocation of retired pay on the member's rank and years of creditable service at the time of divorce. Rather, by not specifically addressing this issue, the USFSPA by implication permits State courts to base the allocation on post-divorce promotions and years of service. A member's pay can increase in the following four ways after the divorce: (1) promotion, (2) longevity, (3) COLA increases, and (4) targeted increases aimed at aligning military pay with comparable civilian sector pay.

In private sector retirement plans, such as a 401(k) plan, the participant's vested account balance or accrued benefit can be valued and divided at the time of separation or divorce. The military retirement system, however, is unlike any private sector retirement plan. The member makes no contribution to the plan, the member has no vested interest in the plan until he or she becomes eligible to retire, and even after becoming eligible to retire the member can be divested of retired pay through punitive action based on the member's misconduct. Even after retiring the member can be recalled to active duty, can forfeit retired pay because of misconduct, and can face certain post-retirement employment restrictions.

These unique features make it difficult for courts to value military retired pay at the time of divorce or separation. As a result, State courts typically award a percentage of the member's retired pay as of the date the member retires. In essence, the State court treats the future promotions and longevity pay increases earned by the member after the divorce as a marital asset. This is inconsistent with the treatment of other marital assets - only those assets that exist at the time of divorce or separation are subject to division. Assets that accrue subsequently are the sole property of the party who earned them. Post-divorce promotions and longevity pay increases are to military retired pay (which is a defined benefit plan) what post-divorce accruals and contributions are to private, defined benefit and defined contribution plans.

A recent DOD study agreed that the calculation of benefits should be based on pay and allowances at the time of divorce rather than the time of retirement. In cases where the member is not retired at the time of divorce, courts often award a percentage of the member's retired pay to the former spouse as of the date the member actually retires. In essence, the court treats post-divorce promotions and longevity pay increases earned by the member as marital assets.

This treatment of military retired pay is inconsistent with

the treatment of other marital assets in divorce proceedings - *only those assets that exist at the time of divorce or separation should be subject to division.* Assets that are earned after a divorce are the sole property of the party who earned them.

Even though DOD concluded that the current practice is unfair to divorced military members, nothing has been done to correct the problem. Congress should at a minimum amend the USFSPA to base all awards of military retired pay on the member's rank and time served at the time of divorce. This provision should be exclusively prospective. The pay increases attributable to promotions and additional time served should be the member's separate property. For example, if a future divorce occurs when the member is an O-4 (i.e., Major/Lieutenant Commander) with 14 years of creditable service, the award of military retired pay must be based on that rank and time served. That the member retires as an O-6 (i.e,, Colonel/Captain) with 24 years of service is irrelevant to the award of military retired pay as property.

An example from a Master Chief Petty Officer in the Navy:

"My Ex-spouse filed for divorce and left me with two minor children three days prior to transferring overseas. Three weeks after our final divorce, she remarried and has remarried twice since. *When we divorced, I was a First Class Petty Officer* (E-6), however she will benefit fourteen years later by receiving 43% of my retainer pay as a *Master Chief Petty Officer* (E-9). No matter how you look at that scenario, it is wrong."

So - why do former spouses continue to receive "windfalls" based upon a military member's continued service and promotions *after* the divorce? There is NO WAY this can be justified!

The following graph shows the difference (windfall) between the division of retired pay based upon what an E-5 with ten years of service is earning at the time of divorce, and what he/she earns at the time of retirement as an E-9 (twenty years after the divorce). Even if it were true that the spouse made an "equal contribution" *during* the marriage, in this example he or she is being rewarded for the four promotions and twenty years of longevity raises which occurred *afterwards*. What contribution was made during *that* time?

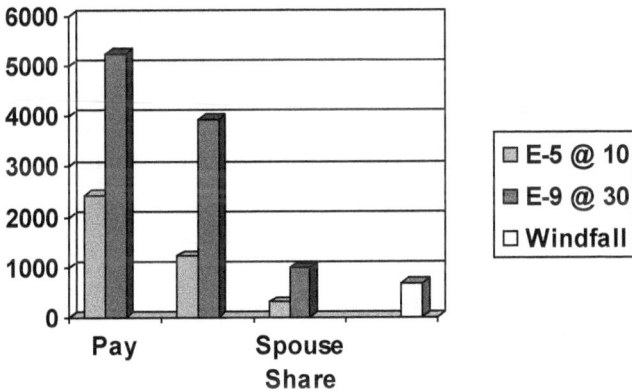

RETROACTIVITY
You Retired When?

USFSPA applies to marriages, and to military careers, which began LONG before the law was enacted. Common sense tells us that it's unfair to change the rules in the middle of ANY game! Why should service members who enlisted and married prior to 1983 (remember, the USFSPA was enacted in 1983, but backdated to 1981 to circumvent *McCarty*) be subjected to a law that did not exist at the time?

The USFSPA had no grandfather clause, so this deficiency changed the military retirement compensation system *retroactively*. Military members already retired, or eligible for retirement, were caught off guard by a law they had no reason to anticipate. The failure to grandfather denied equal property protection under the law to divorced military veterans, whose lives in retirement were devastated without prior notice and without compensatory relief. The USFSPA does, in fact, constitute the "unjust taking" in violation of rights provided by the Fifth Amendment to the U.S. Constitution.

For example, a Navy Chief Petty Officer, who was a Vietnam Veteran, was divorced in 1978 (five years *before* USFSPA was passed) after eleven years of marriage. At the time of divorce, real property was awarded along with child support. In 1991 the military member retired as a Master Chief Petty Officer with thirty years of service. At that time, *the former spouse returned to court* for division of retired pay and was awarded 28% payable <u>at the E-9 rate</u>. At the time of this award, the former spouse was married to her *fourth* husband.

Even in the event that a state court order *illegally* divided military retirement pay before June 26, 1981, in conformity with the FSPA, the order will be honored.

A 1990 amendment to the law addressed the retroactivity problem of former spouses returning to state court on "old" divorces seeking division of the military pension. The FSPA now specifically provides that a state court may not treat retired pay as property in any proceeding to divide or partition it if a final decree of dissolution, annulment, or legal separation (including property settlement cases) was issued before June 25, 1981, and that decree did not divide or reserve jurisdiction to divide the member's retired pay – but that is little consolation to those who have already been victimized!

43

STATUTE OF LIMITATIONS
"Oh, by the way..."

The USFSPA does not stipulate any limits on the time permitted for an ex-spouse to enter a claim for payments. Consequently, the divorced military member can be doomed to a life in retirement waiting for "the other shoe to drop."

With regard to statute of limitations... a time period by which final decrees of divorce may be reopened to divide retired pay as property... the civilian federal plans, again, are statutorily clear... but the Uniformed Services Former Spouses' Protection Act *does not grant such protections.* The courts are currently reopening final decrees, sometimes up to ten or fifteen years after the divorce, and awarding a share of retired pay as property, to include retroactive payments, to long divorced former spouses. This leads to post divorce litigation and legal expenses, as well as virtually ensuring financial ruin for the retiree who has not planned on such an event.

It is time to fix this flaw in the law!

REMARRIAGE
A Cottage Industry

Unlike most alimony, and the annuties of other government agencies, awards made under USFSPA survive the remarriage of a former spouse. As a result, spouses who divorce multiple military members are entitled to collect from *each* of them!

Think that's not possible? A female civilian is presently in her *fourth* marriage to a retired military member. Believe it or not, she is currently collecting USFSPA payments from the *first three* ex-spouses!

The present treatment of military retired pay is inconsistent with treatment afforded federal civilian retirees under Central Intelligence Agency (CIA) and Foreign Service laws. The laws governing how military retired pay is treated in divorce cases are not even remotely comparable to those federal civilian plans, and should, at the very least, be consistent with Foreign Service and the Central Intelligence Agency, whose duty assignments have been likened by

Congress to be very similar. By altering the scheme under which military retired pay is divided in divorce, bringing military retired pay in line with CIA and Foreign Service, the problems inherent within the broad decision making abilities of the states will be eliminated.

There is simply no reason why we should treat men and women who serve honorably in our Uniformed Services any less favorably than men and women who serve in the Foreign Service or CIA.

During the floor debate of the USFSPA in 1982, Congressman Bill Nichols stated:

"The recent change in the Foreign Service Personnel System and the changes proposed in the CIA personnel system currently in conference, have been used to argue the need for change in the military (system).

Both of these systems restrict payments of portions of retired pay to situations in which the marriage lasted ten years, during which time the member was an employee of the system.

In addition, payments of retired pay to former spouses *terminate upon remarriage of a former spouse...*

A similar restriction is only fair for the military member as well, who, to a large extent is similarly situated."

This statement was true then, and it is true now! Why hasn't this flaw been fixed?

DFAS
Should Stand for "Divorced Fools Are Screwed"

There are many documented cases where payments continue to be made to the *new* spouses of *deceased* former spouses. Can you imagine having money taken from the retired pay you earned being sent to the new spouse of your deceased ex-wife or ex-husband?

The Defense Finance Accounting Service (DFAS) requires military *widows* to contact DFAS once a year to verify their eligibility to receive widow benefits. If a military widow is collecting pension payments from DFAS and remarries, then the payments must stop or the widow becomes guilty of fraud!

There is NO REQUIREMENT for former spouses to contact DFAS regarding their marital status. In fact, if your former spouse remarries and has her or his portion of YOUR military retirement check deposited into a joint checking account with his or her new spouse, there is NO LEGAL REQUIREMENT from DFAS for the new spouse to contact DFAS in the event of your former spouse's death.

In other words... if you have no contact with your former spouse and are unaware of their whereabouts or situation, hypothetically they could be dead, and DFAS would still be dividing your retirement pay and sending it via electronic deposit to a joint checking account. The other spouse could be collecting up to fifty percent of your monthly military retirement checks - and your former spouse is dead!

DFAS requires YOU - the military retiree - to notify them if your former spouse is deceased in order to stop division of your military retirement pay. *You* must send DFAS proof of death. For example... a certified copy of the death certificate.

That is not the only problem at DFAS. They are required to accept a court order which is "regular on its face" - which could lead to some unforeseen (and unfair) scenarios.

From a retired Navy Chief Petty Officer:

"During my divorce I discovered my Filipino ex-wife committed bigamy when she married me. In spite of the evidence of her having a prior marriage her requested divorce was granted along with half of my military retired pay. I have since the divorce obtained an annulment of the marriage in the Philippines where the marriage was performed. The marriage document is now amended to read null and void from the beginning of marriage. I sent this to DFAS hoping to have my retired pay fully restored. After two appeals my request remains denied. DFAS says they have a facially acceptable divorce decree and they have no authority under provisions of USFSPA to accept Philippine court orders. Today my ex wife is happy living together with the man she was legally married to before she married me. I have certified copies of both marriage contracts and proof that the first marriage was never lawfully terminated."

SBP
The Ex-Spouse Benefit Plan?

The Survivor Benefit Plan is a plan whereby a portion of a military retiree's pay is paid to a named beneficiary after his or her death. Without this plan, all retired pay would stop when the retiree dies (because it is *not* property!).

The plan, which is partially funded by the government, is paid for by monthly deductions from the retired pay of the member. The amount of the premium is dependent upon the percentage of the retired pay that will be paid to the beneficiary. The maximum amount of coverage is for 55% of the member's gross retired pay.

The decision to elect in or out of the coverage is made at the time of retirement, and is generally irrevocable. THAT MEANS THAT A DIVORCED RETIREE HAS A PREMIUM DEDUCTED FROM HIS OR HER RETIRED PAY EACH MONTH TO FUND AN INSURANCE POLICY FOR AN EX-SPOUSE, AND CAN DO

NOTHING TO STOP IT!

A spouse loses eligibility as an SBP beneficiary upon divorce, although the Department of Defense Authorization Act of 1984 amended the SBP provisions to authorize coverage for former spouses in cases where a court ordered that coverage be established.

When coverage is ordered by a court, and the member then fails or refuses to make the required election, that member shall be deemed to have made such an election if the service finance center receives a written request from the former spouse asking that the election be made.

If the former spouse remarries before the age of 55, coverage is suspended. If the subsequent marriage is terminated by death or divorce, coverage is resumed. As long as the former spouse is alive, the member may not name a current spouse as a beneficiary unless the former spouse waives the benefit in writing.

Current SBP law provides as follows: (1) a current or former spouse who remarries before reaching age 55 automatically loses entitlement to SBP coverage based on the former marriage; (2) a member can designate only one SBP beneficiary; and (3) former members have SBP premiums for current or former spouses deducted from disposable retired pay. In addition, if a member who is required to make an SBP election fails or refuses to do so, the member is "deemed" to have made the election if Defense Finance and Accounting Service (DFAS) receives both a written request from the former spouse and a copy of the court order within one year of the date of the decree of divorce or filing.

Since SBP coverage also stops if the current (surviving) spouse of a retired member remarries before age 55, no compelling reason exists for treating a former spouse more favorably than a current spouse.

The limit on SBP beneficiaries inappropriately deprives the surviving *current* spouse of an interest in the SBP and overcompensates the surviving former spouse. As a result, SBP annuity payments should be divisible or assignable among multiple beneficiaries. Additionally, they should be presumptively divisible in pro rata shares corresponding to the division of the underlying retirement benefits. If the USFSPA is amended to permit designation of multiple beneficiaries, the costs to the Government must be considered.

The rules regarding payment of SBP premiums have led to inequities by requiring that some members and former members pay premiums on annuities for the benefit of former spouses.

In 1987, Congress amended the SBP to permit State courts to order members to participate in the SBP and to designate a former spouse as a beneficiary incident to a divorce. If a member fails to make a court-ordered election, the former spouse may make a "deemed election" of SBP by providing written notice to DFAS within one year of the date of the court order or filing. The SBP benefit is payable to only one beneficiary. It cannot be divided between a spouse and a former spouse or between a child and a former spouse.

THE THRIFT SAVINGS PLAN
A Different Story

The Thrift Savings Plan (TSP) is a Federal Government-sponsored retirement savings and investment plan similar to a civilian 401K. Because it *is* property, unlike retainer pay, *it can and should be divided* as a marital asset during a divorce.

Congress established the TSP in the Federal Employees' Retirement System Act of 1986. The purpose of the TSP is to provide retirement income.

On October 30, 2000, the Floyd D. Spence National Defense Authorization Act for Fiscal Year 2001 (Public Law 106-398) was signed into law. One provision of the law extended participation in the TSP, which was originally only for Federal civilian employees, to members of the uniformed services.

The TSP is a defined contribution plan. The retirement income that you receive from your TSP account will depend on how much you have contributed to your account during your working years and the earnings on those contributions.

What's Mine is Mine, What's Yours is Mine

The TSP offers the same type of savings and tax benefits that many private corporations offer their employees under "401(k)" plans. TSP regulations are published in title 5 of the Code of Federal Regulations, Parts 1600 — 1690, and are periodically supplemented and amended in the Federal Register.

Your TSP account is subject to certain matrimonial court orders and enforcement of your legal obligations to make alimony and child support payments, and to satisfy judgments against you for child abuse. Matrimonial court orders are court decrees of divorce, annulment, or legal separation, or the terms of court-approved property settlements incident to any court decree of divorce, annulment, or legal separation. In order to be considered qualifying and thus enforceable against the TSP, the order must meet the requirements stated in Board regulations (5 C.F.R. Part 1653).

If the TSP receives a document which purports to be a qualifying order or legal process for the enforcement of back payment of alimony, child support, or a judgment against you for child abuse, your account will be frozen for loans and withdrawals. In order to authorize payment from your account, a qualifying court order must clearly identify your TSP account and must describe the award to be made to your spouse, former spouse, or other party in such a way that the amount of the award can be definitively calculated.

A TSP annuity is a benefit paid each month to you (or to your survivor) for life. The TSP purchases the annuity on your behalf from its annuity provider, Metropolitan Life Insurance Company (MetLife). An annuity can only be purchased for $3,500 or more. Therefore, your account balance must be at least $3,500 at the time the TSP uses it to purchase an annuity; or, if you are using only a portion of your account balance to purchase an annuity, that portion must be at least $3,500.

The TSP offers three basic types of annuities:

➤ **Single life** - An annuity paid only to you during your lifetime.

➤ **Joint life with spouse** - An annuity paid to you while you and your spouse are alive. When either of you dies, an annuity will be paid to the survivor for the rest of his or her life.

➤ **Joint life with someone other than your spouse** - An annuity paid to you while you and a person chosen by you (but other than your spouse) are alive. This person must have an insurable interest in you. (See "Choosing a Joint Annuitant Who Is Not Your Spouse" in the booklet TSP Annuities.) When either of you dies, an annuity will be paid to the survivor for life.

Joint life annuities may provide either a 100 percent or a 50 percent survivor benefit. This means that monthly payments will either continue in the same amount (100 percent) or be reduced by half (50 percent) for you or your joint annuitant when either one of you dies.

Several annuity features can be combined with the basic annuity types. Those features are increasing payments, cash refund, and a 10-year certain payout:

With increasing payments, the amount of the monthly payment may increase up to three percent each year, depending on the change in the consumer price index.

With a cash refund, if you (and your joint annuitant) die before receiving payments equal to the amount of the account balance used to purchase the annuity, your designated beneficiary will receive a cash refund equal to the difference between the sum of the payments already made and the annuity purchase price.

With a 10-year certain payout, you receive annuity payments for as long as you live. However, if you die within ten years of the start of your annuity, your beneficiary will receive the payments for the remaining portion of the 10-year period.

If you die before your TSP account is completely withdrawn, the remaining balance will be distributed according to your most recent valid Designation of Beneficiary (Form TSP-U-3), if you completed one. If you did not file Form TSP-U-3, your account will be distributed according to the order of precedence required by law.

Your beneficiaries are entitled to your entire account balance after you die. However, if you die after the TSP purchases an annuity for you, your benefits will be provided according to the annuity option that you selected. If you die while you are receiving your account balance in a series of monthly payments, your beneficiaries will receive the balance of your account in a final single payment.

Payments made directly to spouses of deceased participants are subject to twenty percent mandatory Federal income tax withholding on all tax-deferred money that is being distributed. However, spouses of deceased participants can avoid the mandatory withholding and defer paying taxes on all or part of these payments by having the TSP transfer that amount to a traditional IRA or eligible employer plan (including the spouse beneficiary's existing TSP account). To have the benefit payment transferred to a traditional IRA or plan, the spouse and the IRA or plan administrator must complete Form TSP-U-13-S, Spouse's Election of Payment Method.

THE GENDER ISSUE
Good for the Goose, Good for the Gander?

With women joining the military in ever-increasing numbers, the USFSPA can no longer be considered an issue of husbands providing for wives!

Although the USFSPA's congressional sponsors claimed that the law applies equally to both male and female military members, the only pronoun in the law itself is "his." This may partially explain why the prestigious Defense Advisory Council on Women in the Service (DACOWITS) steadfastly maintained that the USFSPA is of no consequence to women in the service and is, therefore, rarely on the DACOWITS agenda.

When USFSPA was first passed back in the 1980s, the military was still considered to be primarily a male organization. The law was designed to provide for ex-*wives* who were homemakers, had no job skills, and faithfully followed their husbands around from base to base. That has obviously changed.

One of the groups which has supported USFSPA from the

outset is the National Organization for Women, but now that more and more women find themselves in the role of military sponsor it is time for them to reevaluate that stance. Despite what some people would have you believe, this is no longer a gender issue!

The following example is offered: "I am a Chief Warrant Officer in the U.S. Coast Guard. In 1998 my husband of thirteen years divorced me. At the time I was serving aboard a High Endurance Coast Guard Cutter home ported in Seattle, WA. We had been married thirteen years, twelve of which I was in the USCG. I was very much surprised when my spouse informed me that he was in love with another woman. He stated because I spent so much time out at sea that he was lonely and wanted a divorce. During our marriage I had worked two part-time jobs to help put my ex-spouse through college. He received his degree, and was hired by the State of Washington (since 1996). He now has a secure and high paying job with full benefits. We did not have any children, so it was not an issue of support. He is living very comfortably and most likely makes much more than I do. He has since re-married and has a child. As a result of the divorce my ex-husband was awarded 35 percent of my military retirement at my highest rank held. Since the time of the divorce I have advanced in my career significantly by being promoted from E-7 (Chief Petty Officer) to CWO2 (Warrant Officer). The opportunity that I may excel even further before my intended retirement only makes me more angry, knowing that my ex-husband benefits financially as I progress professionally."

It is quite ironic that a law primarily designed to protect the interests of women now causes them to pay a form of permanent alimony to able-bodied *male* former spouses. How can women's organizations continue to support it?

GARNISHMENT
It's Not For Everyone

Retired military members routinely have their pay garnished to satisfy the requirements of USFSPA, but amazingly former spouses *cannot* be garnished in the event of an overpayment which needs to be recouped.

That is not the only flaw in the way garnishments are handled. Because there is literally no oversight, and since application of the law varies from state to state, some courts will circumvent the prohibition against garnishing or dividing VA disability pay as property by ordering it paid as "alimony." Re-characterizing an award of retired pay as alimony, while enabling the former spouse to receive the amount awarded, is inconsistent with the general prohibition against garnishing or dividing VA disability pay – but nobody is doing anything to stop it!

MILITARY RETENTION
Why Stay For <u>Half</u> a Loaf?

Why would a divorced member of the Armed Forces continue towards the goal of retirement when he or she knows a large portion of their retained pay will be given to a former spouse? It makes more financial sense to get out and start a career on the outside!

The USFSPA actually *does* impact national security interests. It adversely affects recruiting, retention, and morale, and puts green replacements into the force in the place of troops who have left in disgust rather than finish a career which gives away their hard-earned retirement benefits.

What if American men and women no longer find it a worthwhile endeavor to join, and make a career of, the United States Military? What if our national leadership, our Congress and the indifference of the American people have sabotaged the only organization dedicated to defending our freedoms... and our very way of life?

The best recruiting poster in existence is a proud veteran, but many military retirees no longer recommend military service to young Americans as a career because of the impact USFSPA has had upon their lives and careers.

The Uniformed Services Former Spouses Protection Act sounds benevolent enough; the financial protection of former spouses of military retirees is its pretended intent. What it has become is the single most reprehensible means of depriving the military retired of the "pension" he or she was promised.

"What," you ask, "is the big problem?" Think of it: with a divorce rate of greater than fifty percent in this country (and higher among our nation's military), it is a given that the married career military service member stands less than a fifty-fifty chance of keeping their earned military retired pay. Who in their right mind will make a career of the armed forces when they realize this great lie has been perpetrated upon those who have gone before them?

As Supreme Court Justice Harry A. BLACKMUN wisely wrote when he delivered the majority 6-3 opinion of the Court on McCarty v. McCarty June 26, 1981.

"(b) Moreover, the application of community property principles to military retired pay *threatens grave harm* to 'clear and substantial' [453 U.S. 210, 211] *federal interests.* Thus, the community property division of retired pay, by reducing the amounts that Congress has determined are necessary for the retired member, has the potential to frustrate the congressional objective of providing for the retired service member. In addition, such a division has the potential to *interfere with the congressional goals of having the military retirement system serve as an inducement for enlistment and re-enlistment* and as an encouragement to orderly promotion and a youthful military."

THE MILITARY DIVORCE RATE
Stand By Your Man (Or Not)

I'M <u>OUT</u>
OF HERE!

Dear John...

In spite of what Doris Mozley said about how USFSPA "helps to keep families together," the existence of this law actually makes divorce an easier and more attractive option. As a result, when the going gets tough (i.e. deployments, war, etc.) the spouses get going!

Between 2001 and 2004 divorces among active-duty Army officers and enlisted personnel nearly doubled, from 5,658 to 10,477, even though total troop strength remained stable. In 2002, the divorce rate among married officers was 1.9 percent - 1,060 divorces out of 54,542 marriages; by 2004, the rate had tripled to 6 percent, with 3,325 divorces out of 55,550 marriages.

More than 3,300 Army officer marriages ended in divorce last year, up 78 percent from a year earlier and triple the number in year 2000. Among Army enlisted soldiers, more than 7,100 were divorced last year, an increase of 28 percent over 2003 and 53 percent since 2000.

Based on data compiled last year, the Defense Finance and Accounting Service had been receiving about 18,000 court orders a year directing the division of military retired pay as part of a divorce settlement.

A large percentage of those divorces are the simply the result of military personnel doing their job, and their spouses decision not to stand by them while they do so. The service member should not be penalized for that! But don't take my word for it - the following letter to *Dear Abby* appeared in the paper on August 30, 2005:

Dear Abby,

My sister needs help. Her husband, "Dale," who has been in the Reserve for 15 years, is being deployed to Kuwait next month, and she's a mess. She went to the emergency room this morning because she thought she was having a heart attack. It was an anxiety attack. One minute she's distraught because he's leaving; the next she wants to divorce him.

"Andrea" was always proud of Dale's service. She has happily bragged that she's an officer's wife, about the pay, the retirement that will come their way, and the travel deals they have enjoyed staying at Army properties all over the country. Until now, she has supported the action in Iraq and Afghanistan.

Now, however, she has kicked Dale out of the house because she believes he has chosen the Army over their family. She says he won't be allowed to call or email her or their two kids while he's on active duty.

Andrea refuses any suggestion of support services through the Army because she doesn't think the session will be kept confidential. Although I want to support her, I believe she's denying Dale the support he deserves. It infuriates me that she has been in favor of the military action as long as it involved other people's families and not her own.

Andrea and Dale have been married for 20 years. She has never lived alone, nor does she have the means to support herself. She has been seeing a therapist for the past few months for depression, but her next session isn't for a few weeks. How can I help? What can anyone else do to help?

- (signed) Concerned Sister

Dear Concerned,

...I hope your sister comes to her senses before it's too late, or she may spend the rest of her life regretting her immaturity and self-centeredness. Her attempts to punish her husband are counterproductive and could sabotage his peace of mind and safety. This is not a matter of choice. Her husband is fulfilling an obligation.

The case of "Andrea" is not an isolated aberration, but is instead an all too common reaction by a spouse to a service member doing his or her duty. The following appeared in the paper on October 13, 2005:

Dear Abby,

Boy, did I identify with the letter about the Army wife whose husband is being deployed to Kuwait. My husband of 25 years is in Iraq now. It's a short deployment; however, my reaction to it was unusual for me.

At first, I took it in stride. But as the time approached for him to leave, I became anxious and depressed. I consulted a therapist, whom I'm still seeing. I had similar feelings as the wife's. I was scared out of my mind that my husband would not return and I, too, wanted a divorce. I'm still mystified about my reaction. He has been away before, but never in a place so dangerous. I, too, felt he was choosing the military over me.

- Alice in Somerdale, N.J.

Dear Alice: **That letter struck a chord with many military (and former military) wives.**

Dear Abby,

What the military wife needs to know is that her reaction to her husband's deployment is not uncommon. She is going through the anger/detachment withdrawal stages - anger at the military and at her spouse for being in the military. It is common to withdraw and/or argue just prior to deployment, since it can be easier to be angry than to confront the pain and loss of departure. She's not the first military spouse to have these feelings.

- Kathie Hightower, Tacoma WA

The following is compiled from articles about the crew of *USS Abraham Lincoln* which appeared in the *Seattle Post-Intelligencer* and *USA Today* in 2003:

After almost eight months at sea *USS Abraham Lincoln* was heading home in December, with crewmen thinking of the warm arms of sweethearts and the tiny hands of toddlers, when orders came to turn it around on January 1.

Morale took a hard hit and stress set in, exacerbated by financial pressures and bad-news messages from home: reports of illness, death, unpaid bills, marital tensions - and "Dear Johns" and "Dear Marys" from disloyal sweethearts.

"They say distance makes a heart grow fonder - apparently not in all cases," Petty Officer Nick Decker said. His buddy, he says, got an e-mail from his wife after they were at sea for two months. She wanted a divorce.

"They have a 1 1/2-year-old," said Decker, who's thankful he's single. "He signed up so that he could support his family. And now she's divorcing him because he left them."

After 9½ months and a war later, the men and women on this floating steel hulk just want off. Homecoming will be a proud moment for much of the crew - but it will also be a bittersweet reminder of the hardships of Navy life. *Lincoln's* crew and aviation wing helped win the Iraq war and didn't lose a single sailor or jet. But they missed more graduations, anniversaries, birthdays and holidays than they care to think about.

Relatives and friends died. Boyfriends and girlfriends found new loves. Kids took their first steps, spoke their first words, went on their first dates. Life at home marched relentlessly on while the sailors and aviators on the world's biggest warship were stuck at sea longer than any other U.S. carrier in the past 30 years.

E-mail, cheaper phones and an occasional videoconference with family kept sailors abreast of home to a degree only dreamed of by veterans of Desert Storm, the 1991 Persian Gulf War. The Navy has wised up about preparing sailors for homecoming since the dysfunction brought on by the Vietnam War's jungle-to-living-room-in-three-days model. But divorce, broken relationships, cheating spouses, and alcohol and drug abuse remain facts of life in communities of returning sailors.

Lincoln's 5,500 sailors and aviators, 10% of them women, were off the coast of Australia, headed home in January when they were ordered to turn around and go back to the

Persian Gulf. A long-anticipated homecoming had been jerked away. Compounding the frustration: No one knew how long Operation Iraqi Freedom would last.

When *Lincoln* finally did arrive in port the crew was greeted by marching bands, Navy Brass, - and some 1500 petitions for divorce.

At the time it was being considered by Congress, advocates of USFSPA described the law as protection for "throwaway wives" who were being abandoned in favor of other women. While it is abundantly clear that is often *not* the case, it is also clear that *all* career veterans are being penalized for the actions of a few.

It is time to repeal or reform this archaic and unfair law!

WHO'S ON FIRST?
Let's Make a Federal (or State) Issue Out of It

It's a STATE issue!

No, it's a FEDERAL issue!

Abbott & Costello may as well be in charge of USFSPA reform!

Courts from time to time weigh legal challenges to USFSPA but typically dismiss them, suggesting it's the job of Congress to modify the law. The legislative branch, on the other hand, typically does not get involved with laws which are already on the books, but instead says it is up to the judicial branch to interpret them.

USFSPA is a perfect example of how the two branches "pass the buck." State divorce courts tell veterans they have no choice but to abide by federal law (USFSPA) when dividing marital "property," and the federal government says they cannot interfere with how individual states handle divorce cases. Talk about avoiding the issue!

In this most recent court case challenging USFSPA, lawyers for the plaintiffs contend the judge ruled illogically when he said service members and retirees can and should raise all relevant constitutional challenges in state court at

67

the time of their divorce settlements. The problem is weaknesses in the law's procedural protections aren't experienced - and therefore "ripe" for litigation - until months or years later when the Defense Finance and Accounting Service (DFAS) actually divides retired pay.

Judge Cacheris, in his 15-page opinion, said federal courts lack subject matter jurisdiction to hear the case (even though it's a *federal* law!), and the plaintiffs lacked legal standing to bring their lawsuit. He also ruled that, because the legal challenges could have been raised in state courts when the original divorce and property issues were settled, a legal principle of *res judicata* applies. The Latin phrase means "the thing has been judged," so a new case is useless.

If that is so, and the principle of *res judicata* applies, does that mean *McCarty v. McCarty* (having already been judged) is still the law of the land, and retainer pay is not divisible? At some point something has got to give!

FORCED RETIREMENT
Well, You <u>Could</u> Retire Now!

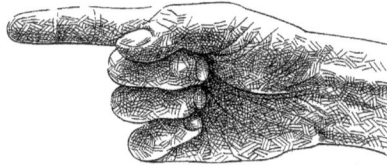

Nothing in the USFSPA allows a State court to order a member to apply for retirement, or to retire at a particular time in order to initiate payments to a former spouse. In fact, the USFSPA contains an *express prohibition* against such action by State courts. Congress and DoD believe such actions would be contrary to the best interests of the United States, and that control of service members must remain with the Federal Government. Nevertheless, State courts have, in some instances, required a member who chooses to remain in military service to make payments directly to the former spouse in an amount equal to what the former spouse *would* receive if the member retired when first eligible. That action constructively *forces* the member to retire!

The USFSPA does not authorize State courts to issue orders that compel the member to retire in order to make retired pay available for a former spouse. To provide for our national defense, the armed forces must be allowed to control when a member is permitted to retire. If military retired pay is awarded solely as property, a court should not be able to compel the member to provide any payments to the former spouse before the member retires. Since the *member* is not entitled to receive retired pay prior to retirement, the *former spouse* should also be precluded from receiving it (when it reflects an award as property) prior to the member's actual retirement.

DoD recommended in 1999 that the USFSPA be amended to explicitly prohibit a court from requiring a member to begin payments (as property) to a former spouse before actual retirement, as economically, this may compel the member to retire – but of course no action was ever taken to eliminate this loophole.

Such an order directing payments to commence upon reaching twenty years of service was recently issued in the case of Lieutenant Colonel Patricia Larrabee, and actually forced her to retire prematurely. She recently told Secretary of Defense Rumsfeld, "I can't afford to write (a) check to my ex-husband every month out of my military pay," during a 2005 forum televised worldwide to U.S. troops over the Pentagon Channel. "By the way," Larrabee added, "he makes thousands and thousands of dollars more than I do." Larrabee told Rumsfeld she had custody of the couple's two young children, and that her ex-husband had "resigned from the military because it wasn't lucrative enough for him." During their nine-year marriage, she said, "he tripled his income due to the support I provided him while he went to school full time. And by the way, I supported the family with my military paycheck."

"Sir," Larrabee told Rumsfeld, "we are your supporters - some of your biggest supporters in this country - and we would like to get support from our leadership as well."

This begs the question: How can someone be required to make payments based upon something they do not have, and to which they themselves are not yet entitled?

THE OTHER BENEFITS
Medical, Commissary and the Exchange

If former spouses "earn" a portion of a service member's retainer pay simply by being married to the member (with no 20 year requirement), why then are they not also entitled to full exchange, commissary and medical privileges? And why do former spouses who are entitled to those perks lose them in the event they remarry, while retaining their entitlement to the member's retainer pay? Because it comes out of the *governments* pocket, rather than the member's – *that's* why!

There are the rules for the "other benefits":

Full Privileges - the "20/20/20" Former Spouse:

Full benefits (medical, commissary, base exchange, theater, etc.) are extended to an *unremarried* former spouse when:

1. The parties had been married for at least 20 years.
2. The member performed at least 20 years of service creditable for retired pay; and
3. There was at least a 20 year overlap of the marriage and the military service.

71

Concerning medical care, if the former spouse is covered by an employer-sponsored health care plan, medical care is not authorized. However, when the former spouse is no longer covered by the employer-sponsored plan, military medical care benefits may be reinstated upon application by the former spouse.

If a 20/20/20 former spouse remarries, eligibility for the benefits are terminated. If the subsequent marriage is ended by divorce or death, commissary, base exchange, and theater privileges may be reinstated. Medical care cannot be reinstated.

Limited privileges - the "20/20/15" Former Spouse:

For divorces before April 1, 1985, a four year renewable identification card authorizing medical benefits (no commissary, base exchange, or theater privileges) is awarded to an unremarried former spouse when:

1. The parties had been married for at least 20 years.
2. The member performed at least 20 years of service creditable for retired pay; and
3. There was at least a 15 year overlap of the marriage and the military service.

Concerning medical care, if the former spouse is covered by an employer-sponsored health care plan, medical care is not authorized. However, when the former spouse is no longer covered by the employer-sponsored plan, military medical care benefits may be reinstated.

Divorces on or after April 1, 1985 and before September 30, 1988:

These 20/20/15 former spouses qualify for medical

benefits for two years from the date of the divorce, dissolution, or annulment or December 31, 1988, whichever is later. If the former spouse is covered by an employer-sponsored health care plan, medical care is not authorized. When the former spouse is no longer covered by the employer-sponsored plan, military medical care benefits may be reinstated. However, any reinstatement may not extend beyond the original two year entitlement.

Divorces on or After September 30, 1988:

These 20/20/15 former spouses qualify for medical benefits for one year from the date of the divorce, dissolution or annulment. If the former spouse is covered by an employer-sponsored health care plan, medical care is not authorized. When the former spouse is no longer covered by the employer-sponsored plan, military medical care benefits may be reinstated. However, any reinstatement cannot extend beyond the original one year entitlement.

Former Spouses who were not at least "20/20/15" spouses do not qualify for any entitlements! *So why is it they are entitled to a portion of a servicemember's retainer pay?*

Private Health Insurance:

From time to time private insurance programs have been established to provide transition coverage for former spouses who will lose medical coverage. The most recent program was established in 1994, the CONTINUED HEALTH CARE BENEFIT PROGRAM (CHCBP). Although the program was designed primarily for those military members who are separated under one of the new incentive programs, coverage for former spouses is available. Former spouses who do not remarry are eligible to purchase coverage for up to 36 months. Information can be obtained from CHCBP at 1-800-809-6119.

Points to Ponder:

1. Statutory right. The privileges granted are a matter of statutory right. There are several implications from this fact. First, there is no discretion given to any government official to expand privileges. If the tests are not met, the privileges do not exist. Period!

2. Effect of Remarriage. Since medical benefits are permanently extinguished upon remarriage, it is imperative that non-member clients be advised of this rule. Practitioners frequently confuse the fact that remarriage may not affect the non-member's right to a share of the retired pay as property to the effect of remarriage on other issues, such as benefits. (Remarriage will also affect survivorship rights if the non-member is covered under the Survivor Benefit Plan. See 10 U.S.C. §1450.)

Commissary and Exchange Privileges:

Commissary. The purpose of the commissary privilege is to make items of convenience and necessity, especially items related to subsistence, available for purchase by military personnel at convenient locations and reasonable prices. The types of merchandise and food items authorized for sale at a commissary are specifically limited by legislation.

Exchange. The purpose of a military exchange is to provide merchandise and necessary services at moderate prices to authorized patrons. An additional purpose is to generate earnings to supplement appropriated funds for the support of DoD's Morale, Welfare, and Recreation (MWR) programs.

There is no specific statutory authority that governs the establishment and operation of military exchanges. Rather, they are established and operated under regulations promulgated by the military departments.

Value of Commissary and Exchange Benefits. The annual savings from using commissary and exchange stores as compared with use of commercial retail stores has been estimated to average between 20 and 25 percent.

General Eligibility Requirements. Several categories of individuals are entitled to use commissary facilities, including active and retired members and their surviving spouses (unlimited use), veterans with service-connected disabilities and their surviving spouses (unlimited use), and certain members of the Reserve Components and their spouses (currently, 24 visits per year). All of these categories of individuals have unlimited use of exchanges.

Eligibility of Former Spouses. *Certain <u>unremarried</u> former spouses* are entitled to commissary and exchange privileges "to the same extent and on the same basis as the surviving spouse of a retired member of the uniformed services." To be entitled to commissary privileges, the following requirements must be satisfied: the former spouse must not be remarried, must have been married to a member who completed at least 20 years of creditable service, must have been married to such member for at least 20 years, and must have a marriage/creditable service overlap of at least 20 years.

BOTTOM LINE: The same rules do not apply to the government with regard to benefits "earned" by a former spouse.

DIVIDING DISABILITY PAY
After All, You've Got <u>Two</u> Legs!

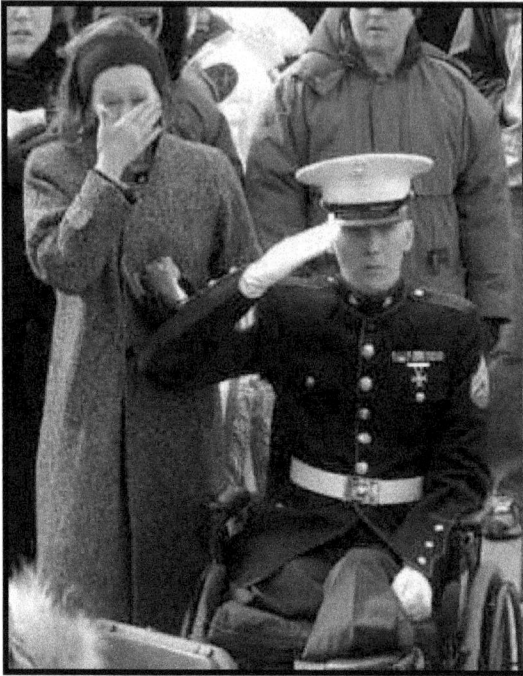

No spouse has ever lost a limb while shopping in the PX!

Former spouses are not content with the division of retainer pay – they are also campaigning for legislation which would allow for the division of *disability* pay.

In *Mansell v. Mansell*, a California court had divided the husband's military *disability* payments - but fortunately the U.S. Supreme Court reversed the decision, holding that USFSPA <u>did not authorize</u> treatment as divisible property of divorce military retirement pay which had been waived so as to receive veteran's disability benefits. Some creative divorce courts have found a way around this however, and instead award some former spouses *permanent alimony* (which naturally must be paid directly to the ex-spouse by the veteran out of his or her disability pay). This practice is beyond the pale.

The USFSPA provides that disability pay being received under Chapter 61 of USC 10 1201 et seq. is not to be included in the calculation of "disposable pay." As previously stated, this decision was reinforced in 1989 by the U.S. Supreme Court's ruling in *Mansel*. Despite this, divorce courts frequently include disability pay in the calculation of disposable pay under the USFSPA. Since USFSPA victims usually do not have the financial resources required to seek remedies for this in the courts, further legislative relief is required to ensure equal protection under the law.

Many service connected disabilities which the Veterans Administration deems service related in nature were received in combat or combat training situations. Typical examples would be back and spinal cord injuries which are common among paratroopers having a "hard" landing. Any military specialty that requires heavy lifting of military hardware on a daily basis, i.e. missiles, rockets, bomb racks, pylons, heavy guns and ammo. Hearing loss from hazardous noise in all branches of the service. Head and neck injuries sustained in wrecks in government vehicles. Many types of cancers and diseases from exposure and handling of biological, chemical agents, and hazardous materials. Post Tramatic Stress Disorder (PTSD) from being a witness of the horrors of war.

How can anyone with an ounce of conscience consider a service connected disability to be a divisible community property asset? These disabilities very often negatively impact the quality of the rest of the retiree's life. A former spouse that did not get injured or sustain any service connected injuries should not be able to receive half of that financial compensation for such injuries. The former spouse is not the one that sustained the injury or disability, whether it be a physical aliment or mental disorder.

While Congress intended for disability retired pay to be income belonging solely to the retiree, and it has NEVER been thought of as deferred income, Federal Statutes allow disability retired pay to be garnished for both child support and alimony, granting the states the ability to creatively manipulate both federal and state laws. By renaming the property division "alimony" the courts sidestep the intent of the Supreme Court in *Mansell v. Mansell* (490 US 581) and are modifying the bonds of the settlement agreement, creating a new form of federal entitlement through case law.

Full examination must be given to all aspects of this issue. Often disability retired pay is the sole source of income for a disabled retiree, and as it stands now it may be divided with a former spouse to pay alimony, child support, arrears, or all three. While federal law mandates that only up to 65% of the retired pay may be garnished, the USFSPA grants any means under law to the States for fulfilling these obligations. Under contract law, the courts can obtain what cannot be garnished through a monetary judgment or real property. This could essentially cause a disabled military retiree to lose up to 100% of what is intended to be old age security.

The Department of Veterans' Affairs does not take lightly the determination of disability for any member of the Armed Forces. Application for disability is scrutinized very carefully and the military veteran must undergo strict physical examinations to determine their eligibility. But groups representing former spouses don't care about any of that. They want every penny they can get, regardless of the sacrifices of the military member.

WITHOUT FAULT, MERIT, OR NEED
USFSPA Horror Stories

Anyone who believes the USFSPA is a fair law designed to protect the interests of "widows and orphans" (as alleged by Doris Mozley) needs to read these three nightmarish examples of fraud and abuse. How can *anyone* sanction such miscarriages of justice?

WITHOUT FAULT: An Air Force Colonel was taken prisoner-of-war, sent to a North Vietnamese POW camp in the Fall of 1967, and later repatriated to the U.S. in 1973. Shortly after returning home after six years in captivity, he was served with divorce papers. According to those papers, the court declared the "date of separation" from his spouse to be April 1, 1970 - while he was still a prisoner-of-war! The former spouse did not have to repay any of the pay and allowances she received and spent AFTER the "date of separation"; she was also entitled to his accrued leave pay, and to monies he received under the War Crimes Act for inhumane treatment. The former spouse was also awarded the home, car, 42.7% of the member's military retired pay, child support and spousal support (even though the former spouse had numerous open affairs during the member's incarceration in a POW camp - and ended up marrying the attorney who prepared the divorce action.

WITHOUT MERIT: A Marine Corps Staff Sergeant, returning to his duty station in Twenty-Nine Palms after serving in combat during Operation Desert Storm, planned to retire with twenty years honorable military service. Upon his arrival home, his wife of nineteen years was found cohabitating with another man. In May 1991, the spouse abandoned the service member and their three children and filed for no-fault divorce in California. The divorce was final in January of 1992. The military member was ordered to give 50% of the property of the marriage to the former spouse, and 47.5% of his military retired pay.

WITHOUT NEED: An Air Force Master Sergeant served twenty years in the military, including two tours in Vietnam. He and his wife were married the final sixteen years of his military service. While stationed in Alaska, and entering his last year before retirement, he was sued for divorce by his wife (who had found a boyfriend) and evicted from his home. The court awarded the ex-spouse 40% of the service member's retired pay as property, and an additional 27% as child support. After taxes, the retired service member receives approximately $130 monthly. Keep in mind the former spouse was employed at $34,000 per year, and her live-in boyfriend was employed at $26,000 per year. Even so, the military retiree pays her $9,000 per year.

For more horror stories visit www.ULSG.org

THE RUMSFELD TOWN HALL MEETING
Say It Ain't So, Don!

During a "Town Hall Meeting" with the troops and the Chairman of the Joint Chiefs in June of 2005, Secretary of Defense Donald Rumsfeld was asked about a provision of USFSPA which compels active duty members to begin making payments to a former spouse once they reach twenty years of service – <u>even if they don't retire!</u> This exchange was particularly enlightening because the question was posed by a *female* Army officer, and the Secretary confessed that he had never even *heard* of USFSPA – and the fact that Mr. Rumsfeld is no longer SecDef in no way diminishes the impact of this exchange.

Q: Sir, this is for you, Mr. Secretary. I'm an active-duty lieutenant colonel, divorced, full custody of two small children. My ex-husband resigned from the military because it wasn't lucrative enough for him.

During our marriage, our nine years together, he tripled his income due to the support I provided him while he went to school full- time. And by the way, I supported a family with my military paycheck.

Now I'm living with a divorce decree that not only directs me to provide a large chunk of my retirement pay to him; it also directs me to start paying him upon reaching twenty years in service, whether I choose to retire at twenty years or not. This is forcing me out of the military next year. I can't afford to write a paycheck - write a check to my ex-husband every month out of my military pay. By the way, he makes thousands and thousands of dollars more than I do.

This is a result of the Uniformed Services Former Spouses' Protection Act. I'm not the only one affected by this injustice. There are many other injustices that have been imposed on military members for years.

Sir, we are your supporters, some of your biggest supporters in this country, and we would like to get support from our leadership as well.

SEC. RUMSFELD: This is a –

Q: And so –

SEC. RUMSFELD: This is a statute, the –

GEN. MYERS: Right. It's a law.

SEC. RUMSFELD: A law.

GEN. MYERS: In the past.

Q: Sir. Yes, sir. Uniformed Services Former Spouses' Protection Act, which, sir, I was told that you supported.

SEC. RUMSFELD: I've never heard of it. (Laughter)

Q: And, sir, as you may know, or may not know, the divorce rate in the military is much higher than it is in the civilian sector, and it is growing. And –

SEC. RUMSFELD: When did this law go into effect?

Q: Oh, sir, people have been trying to fight this for twenty years.

GEN. MYERS: Yes, it's old. It's a couple – it's at least 15, 20 years it's been around, right? Ten, 15, 20 years?

Q: Well, before I came into the military, sir.

GEN. MYERS: Right.

SEC. RUMSFELD: Well, I'll be happy to have David Chu look at it. I'm just not knowledgeable, I'm afraid, about it.

Q: Okay, well –

GEN. MYERS: It was different -- actually, it was created, I think, in different times. I think was part of the mindset when spouses were normally women –

Q: Yes.

GEN. MYERS: -- and when they probably did not work, and when --

Q: But sir, even –

GEN. MYERS: Yeah. So it needs to be looked at. I think the Secretary's idea is a good idea.

Q: May I say one more thing, please, sir? I know that it was set for a much earlier generation. But I will say that since I've been in the military, since August of 1986, everywhere I've been stationed, and Germany included, even female spouses have had opportunities for jobs, given preference for government jobs, had opportunities for education beyond high school. There's always some sort of college program.

So although you may look and this may sound a little bit shocking to you because now there's a woman having to pay an ex-husband who makes just a lot more money than a lot of us in this room, this is an issue that is not a gender issue, it is a military service member issue. And, frankly, we need some support, and we'd like for you to support change or congressional amendment to the current act and actually help promote it, because we can't get a congressman or anybody to touch this.

SEC. RUMSFELD: We'll have David Chu take a look at it. Thank you. (end of exchange)

That was the end of the exchange, and as far as I am aware the end of the investigation. To the best of my knowledge Mr. Chu never researched the issue or briefed the Secretary on the provisions of USFSPA, and if he did it was done without any input from anyone outside of the Pentagon. The following article was written by Retired Army Colonel George Tate in the wake of the Rumsfeld revelation, and raises some interesting questions:

WHAT DID HE KNOW
And When Did He Know It?

By Colonel George W. Tate, US Army (Ret)

Famed for his ability to DETECT tough questions from the media and from Members of Congress, on June 29 Secretary of Defense Donald Rumsfeld was stopped dead in his tracks by a question from one of his own; a lady lieutenant colonel, who brought up a valid and vexing question about military retirement pay. During a DOD "town hall" meeting, aired the world over, she asked him why she faces the prospects of having to pay a large chunk of her retirement pay to her much wealthier ex-husband. Why indeed!

The usually glib and facile Rumsfeld, uncomfortably facing a very public, televised moment of truth he'd rather not face, blinked. Summoning the only dodge he could summon, the Sec Def trotted out his best "Aw shucks" manner and asked, "Is that a statute?" "I'm unaware."

Indeed, Mr. Secretary, "that's a statute." It's 10 USC 1408, the Uniformed Services Former Spouses' Protection Act, and it's been on the books for more than 24 years. And yes, Congress deviously backdated it to avoid an inconvenient conflict with a Supreme Court ruling which held that military retirement pay could not be divided as property.

That statute, Mr. Secretary, is devastating the morale of your troops. Though DOD has hidden the law under the rug for all those years, word of this DOD back-stab is finally reaching the field. In this global war on terror, in this era of back-to-back deployments to battle insidious terrorism, the very troops tasked to accomplish those dirty missions for

America are subject to a discriminatory law mandating that they hand over up to half of their retirement pay permanently to a former spouse. And that's in addition to alimony and child support. Only military members, alone among all Americans, are subject to such a law providing lifetime benefits to ex-spouses above and beyond those available to the public at large and continuing even after the remarriage of the former spouse.

The Secretary's dodge won't wash.

His answer feigning ignorance of the law can be interpreted two ways. Either he has been shielded by his underlings from the grim truth of what the USFSPA is doing to troop morale, in which case he's made to look incompetent, or he took the unwise path of pretending not to know in order to cover his tracks. In other words, he lied. Which of these is the more likely?

In 2001, on Rumsfeld's watch, the Department of Defense provided Congress a voluminous report, delivered two years late to Capitol Hill, describing why the law is good for the troops. Baloney! The report silently, and without comment, overturned the "on-the-Congressional-Record" position of prior administrations regarding the USFSPA.

And as you read this, Rumsfeld's lawyers are scrambling to protect him from a federal lawsuit which names him, by name, as defendant, and which challenges the constitutionality of the law. That case is alive and well in Federal District Court in Alexandria, VA, and if Mr. Rumsfeld wants us to believe that he doesn't know about the USFSPA, he must think we're all pretty stupid. This case is far from resolution, and if Mr. Rumsfeld doesn't know his personal chestnuts are in the fire, we think he's pretty stupid.

Clearly Mr. Rumsfeld does know about the USFSPA. Just as clearly, his tap-dance on camera on June 29 was a disingenuous attempt to avoid public airing of the Department's long-standing support for an ill-conceived law which DOD leaders support at their peril, and which decimates the morale of military career personnel, just when we need them most.

Though DOD's civilian lawyers fabricate fanciful theories for the Department's aggressive and inexplicable support for the USFSPA, in fact the USFSPA created a unique, unprecedented, discriminatory, and wholly unnecessary body of law for military spouses only, on top of their rights comparable to other divorced spouses. The USFSPA may have been well intentioned, but it has harmed far more than it has helped, and now more than 150,000 of America's defenders have had their old-age security cashiered by it. Some have been forced into bankruptcy and all are blindsided in divorce by DOD's dirty little secret. The number of America's finest who have been duped and stabbed in the back by their own chain of command rises daily.

Wake up, Mr. Rumsfeld! Recruiting is down. Divorces are up. Military divorces, until recently less frequent than the general population, now exceed the general rate. American troops are being sent into harm's way with stressful frequency, against an insidious enemy which beheads and car-bombs innocents. Throughout it all, the Department maintains its "head-in-the-sand" attitude about the USFSPA and that law's pernicious effects on troop morale.

Maybe it's even worse than "head-in-the-sand"; maybe the Department shovels the sand over its own head and hides from the unpleasant truths about the USFSPA. It has for more than 24 years failed to provide briefings and orientations to the troops to tell them that they may dedicate a lifetime of honorable, often dangerous service to America only to have their retirement security handed over to a third party civilian who took no oaths and never stood in harm's way. This, remember, is the same DOD which clandestinely reversed the policy position of previous administrations in its 2001 report to Congress.

Even more indicative of DOD sentiment, in the current lawsuit in federal court, Mr. Rumsfeld's lawyers fought (unsuccessfully) to

exclude testimony from a soldier in Iraq. And Mr. Rumsfeld says he didn't know?

Are military families important? Of course they are. Do they deserve more preferential treatment than their military sponsors? What theory could possibly justify *that*? Do military spouses deserve lifetime monetary awards that civilian spouses don't get? Why? Is the playing field level in military divorces? Absolutely not! Why the Sec Def thinks that's a good thing is known only to him.

As the lawsuit progresses, maybe he can explain it to the judge, and to the troops who wonder. And to the courageous lady lieutenant colonel who nailed him on June 29. Until the Congress amends the USFSPA to eliminate its anti-military bias, or until it's declared unconstitutional by the US Supreme Court, it will be the single most demoralizing factor to the career military force. Mr. Rumsfeld knows that, and he knew it on June 29.

KEEPING THE TROOPS INFORMED
Shhhhhh!!!

The military provides briefings, orientations and classes for every subject from sexual harassment to digging a straddle latrine, but make no effort to educate the troops about the possible effect of USFSPA upon their retirement benefits or, for that matter, its very existence. As noted by the Supreme Court, "such a division has the potential to interfere with the congressional goals of having the military retirement system serve as an inducement for enlistment and re-enlistment and as an encouragement to orderly promotion and a youthful military."

The fundamental promise of a 20-year, half-pay retirement benefit to service members should be acknowledged as having been withdrawn by the United States government and should be eliminated from all recruitment materials. Given the high divorce rate in the military, *the provisions of the FSPA should be explained to all members upon enlistment and reenlistment.* Our volunteer soldiers and sailors deserve to know the facts concerning this important area of military compensation so that they can intelligently decide whether to assume the risks and obligations of military life.

Administration of the USFSPA by DoD

The armed forces claim to provide accurate, readily available information on the USFSPA in the form of handouts and fact sheets, which are distributed through legal assistance offices and various other outlets. The armed forces, with the exception of the Marine Corps, generally rely on their judge advocates and civilian attorneys to render advice and assistance with regard to the USFSPA. The Marine Corps has designated its Separation and Retirement Branch as its point of contact for information related to the USFSPA.

The problem is, by the time a soon-to-be-divorced troop shows up in a military attorney's office and learns about USFSPA, it is already too late. Such a policy simply locks the barn door after the horse has gotten away. What the troops really need is a briefing about USFSPA when they *first enter the military*, and a refresher if and when they decide to get married.

It is reasonable to assume no "USFSPA Briefing" is being given because it could put the military in a negative light and serve as a detriment to recruiting, and if that is indeed the case it is nothing short of criminal. The Armed Forces pay a lot of lip service to being a "band of brothers," and while that concept is generally borne out (by the troops) on the battlefield, it seems to have been forgotten in the higher echelons of command. Why not put *all* of the cards on the table, and let those who defend our nation make their own decisions based upon that information? They deserve at *least* that much!

REFORM LEGISLATION
A Political Hot Potato

Several members of Congress have sponsored legislation to rectify some of the glaring deficiencies in USFSPA (i.e. the award of windfalls, termination upon remarriage, etc.) but none of them have made it out of committee...

The Department of Defense "recently" finished a report which was ordered by Congress to be completed and presented by September 30, 1999. This report was finally completed in June of 2002, almost three years late. Its non-completion is the reason many Congressmen and Senators refused to visit previous bills before them to reform the USFSPA. Now that the report is available to Congress, and *even before* the events of September 11th, no one wanted to touch this hot potato. It seems that it's too politically risky to do the right thing, and reform has garnered only lip service from most legislators. Now Congress is sending troops into harm's way to defend our freedom. What kind of welcome will they get when they return? Will their spouses be waiting for them with divorce papers?

Some of the reform bills which have been introduced, but never acted upon, are as follows:

The USFSPA reform bill in the 107th Congress H.R. (1983) had four provisions:

1. Terminate payments of retired pay to remarried former spouses, both prospectively and retroactively.

2. Require that payments to former spouses be based on length of service and pay grade at time of divorce, not time of retirement. This would have applied both retroactively and prospectively. ("Windfall benefit" provision.)

3. Establish a two-year statute of limitations for former spouses to apply for a division of retired pay following divorce. Applied retroactively to June 25, 1981.

4. Absolutely prohibit payments of disability pay to former spouses, including garnishment under the Social Security Act, except for child support.

The USFSPA reform bill in the 108th Congress (H.R. 111) had five provisions:

1. Quantification of the share of retired pay payable to former spouses. Contains a formula based on (i) years of marriage while the member was qualifying for retired pay; (ii) time credited for retired pay. Applies to divorces occurring before or after enactment. Contains an exception for voluntary spousal agreements.

2. Limitation on duration of payments. For divorces

occurring after enactment: (i) Where the time married while the member was on active duty (or, in the case of reserve/National Guard, was earning retirement points) is less than 20 years, payments would be made for the number of years married while the member was qualifying for retired pay, or, until the former spouse remarries, whichever occurs first. (ii) Where the time married while the member was on active duty is 20 years or more, payments would continue until the death of the member or former spouse, whichever occurs first.

For divorces occurring before enactment: (i) If the length of the marriage while the member was on active duty was less than 20 years, payments would continue for the length of the marriage while the member was on active duty. (ii) If payments have already been made for the number of years the member was on active duty, payments continue for two years after enactment and then end. (iii) If the payment period ends within two years of enactment, payments would continue until the end of the two-year period after enactment. (iv) If the length of the marriage while the member was on active duty was 20 years or more, payments would continue until the death of the member or former spouse, whichever occurs first.

3. Windfall benefit. "Windfall benefit" provision, but applying only to court orders issued after date of enactment (prospective application only).

4. Statute of limitations. Two-year statute of limitations, applying only to court orders issued after date of enactment (prospective application only). However, courts are prohibited from ordering payments in arrearages for more than two years.

5. Protection of disability pay. Absolute prohibition against payment of disability pay to former spouses, including garnishment under the Social Security Act of VA disability compensation received in lieu of waived retired pay (except for child support).

UNIFORMED SERVICES DIVORCE EQUITY ACT of 2005 (USDEA-05) This Uniformed Services Former Spouses Reform Act Bill was pending introduction in the 109th Congress. Its five main provisions are outlined as follows.

1. Correlate USFSPA payments with the duration of the active duty marriage:

a. For active duty marriages of less than 240 months, limit the period of payments to the former spouse to the time married, or until remarriage: whichever comes first.

b. For active duty marriages of 240 months or more, payments to a former spouse would be made for life, regardless of their marital status.

2. Eliminate the provision requiring an active duty marriage of 120 months or more before the former spouse can be paid by DFAS

3. Base payments on the service member's pay grade at the time of divorce to eliminate the "windfall benefit".

4. Prohibit payment of imputed income for divorcing active duty service members.

5. In cases where divorce settlements are reopened, limit payments in arrears to two years.

What's Mine is Mine, What's Yours is Mine

The following article appeared in *The Florida Bar Journal* **in December of 1997:**

In the early 1980's Congress enacted legislation overturning the U.S. Supreme Court decision in *McCarty v. McCarty*, 453 U.S. 210 (1981), which held that a military pension was the separate property of the service member and not subject to division in a dissolution of marriage action. In the 15 years since the enactment of the Former Spouse Protection Act (FSPA), controversy surrounding the fairness of this legislation, and the implementation of the FSPA by the various states, has raged in Congress. This article will discuss the McCarty decision and its merits, the FSPA and its several amendments, Florida's approach to division of military pensions, and a brief overview of application of the FSPA in other states.

The *McCarty* Decision

On June 26,1981, the Court held that, in a dissolution of marriage, federal law precluded a California court from dividing military non-disability pay pursuant to state community property laws. The Court found that dividing a military pension in state court threatened grave harm to "clear and substantial" federal interests such as providing for the retired service member in old age, encouraging enlistment and reenlistment, orderly promotions, and encouragement of a youthful military. Military pensions were viewed by the Court as different from other pension systems, because the retired officer is subject to recall to active duty at any time, continues to be subject to the Uniform Code of Military Justice, and is restricted in post-service activities, including employment. Because of these factors, military retirement had not historically been considered a "pension," but rather reduced pay for reduced services. Not even a limited "property" concept had ever existed in military compensation laws or the Court's own precedents. As early as 1881, the U.S. Supreme Court had ruled that when a military member retires or leaves active duty, compensation is continued, with reduced duties and responsibilities. *U.S. v. Tyler,* 105 U.S. 244 (1881).

The *McCarty* Court also noted that dividing military pensions made it less likely that the retired service member would choose to reduce his or her retirement pay still further by purchasing an annuity for the surviving spouse and children. Since the military retirement laws contained nothing permitting the states to divide a military pension in a dissolution of marriage, the California superior court was reversed. Congress was invited to change the law if it so desired.

The Former Spouse Protection Act

Congress did act shortly after the *McCarty* decision by enacting the Uniformed Services Former Spouses' Protection Act. The USFSPA, or FSPA or FSVA, is found at 10 U.S.C. §§1408 et seq. (1982), effective date February 1, 1983, retroactive to June 25, 1981, one day prior to the *McCarty* decision. Sponsored by Rep. Patricia Schroeder (D-CO), the FSPA reversed the *McCarty* decision, rejecting the Court's concerns regarding military retention, enlistment, and the economic needs of older veterans. Also known as the Former Spouse Victim act by military retirees, the FSPA has been a source of confusion and controversy at both the state and national level.

The FSPA applies to the "uniformed services," defined to include the Army, Navy, Air Force, Marine Corps, Coast

Guard, commissioned corps of the National Oceanic and Atmospheric Administration, and the commissioned corps of the Public Health Service. The FSPA applies to active duty, retired, and reserve/guard (whether active duty, inactive status, or retired), pay and nonpay categories.

Since the FSPA is a federal statute, its provisions and the regulations thereunder preempt or supersede state laws. A state court order that contradicts the FSPA will not be enforceable. The FSPA, with limitations, allows state courts to treat a military pension either as property solely of the service member, or as property of the member and his or her spouse in accordance with the law of the jurisdiction for pay periods beginning after June 25, 1981. In the unlikely event that a state court order divided military retirement pay before June 26, 1981, in conformity with the FSPA, the order will be honored.

A 1990 amendment to the law addressed the retroactivity problem of former spouses returning to state court on "old" divorces seeking division of the military pension. The FSPA now specifically provides that a state court may not treat retired pay as property in any proceeding to divide or partition it if a final decree of dissolution, annulment, or legal separation (including property settlement cases) was issued before June 25, 1981, and that decree did not divide or reserve jurisdiction to divide the member's retired pay.

Special Federal Jurisdictional Rules

Special federal jurisdictional rules apply to divide a military pension. A state court may not divide a military pension unless the court has jurisdiction over the member by reason of his or her 1) residence, other than because of

97

military assignment, in the territorial jurisdiction of the court; 2) domicile in the territorial jurisdiction of the court; or 3) consent to the jurisdiction of the court. It is entirely possible for divorce, child support, and alimony jurisdiction, but not pension jurisdiction, to exist in a case where the service member declines to participate and does not seek affirmative relief.

The Ten Year Rule

A state court may not effectively order direct payments of pension benefits from the Defense Finance and Accounting Service unless the former spouse was married to the service member for ten years or more, during which time the member performed at least ten years of service creditable for retirement purposes. A state court is not prohibited by the FSPA from dividing a military pension in a marriage of less than ten years, however direct payment of the pension benefit to the nonmilitary spouse by DFAS is not permitted. There is no ten-year marriage requirement for garnishment from Defense Finance of child support, alimony, or both.

The FSPA also defines the pension benefit that is available for division. Net disposable retirement pay, not gross retirement pay, is subject to division. Subtract from gross retirement pay any debts owed the U.S. government, survivor benefit plan premium (if any), court martial fines and forfeitures, and amounts waived to obtain disability pay. Net disposable retirement pay is the remaining amount. The order dividing retirement benefits must provide specifically for payment of a fixed amount expressed in U.S. dollars or payment as a percentage or fraction of disposable retired pay. Percentage orders automatically award the same percentage of cost of living increases to the former spouse. Certain formula orders are now being accepted by Defense

Finance. Taxes and Social Security obligations of each retirement pay recipient are separately withheld, and separate 1099's are issued to each party for divorce decrees effective after February 2, 1991. The total amount of the disposable retirement pay of a member that may be paid to a former spouse or spouses may not exceed 50 percent.

Downsizing or "right-sizing" of the military forces in the 1990's has resulted in many members receiving voluntary separation incentive or special separation bonuses. See 10 U.S.C. §§1174a, 1175. The Florida Supreme Court has held that these programs are sufficiently similar to retirement benefits that an agreement or court order distributing military retirement may be enforced against these benefits. Kelson v. Kelson, 675 So. 2d 1370 (Fla. 1996). No direct payment regulations have been issued, so enforcement at law is problematic. A former nonmilitary spouse may receive unlimited retirement splits from former military spouses. Although rare, documented instances of former military spouses receiving two or three pension interests exist. The former nonmilitary spouses' retirement benefits are not reduced or eliminated upon a remarriage, as they are in many other federal pension programs. However, payments do terminate on the death of the service member. Therefore, insuring the military pension is an important consideration when representing the nonmilitary spouse.

The regulations also provide that the member's retirement pay may be garnished to satisfy a court order that provides for the division of retired pay as property, not to exceed 50 percent of disposable retired pay for all court orders, or 65 percent of disposable retired pay for all court orders and garnishments for child support and/or alimony. If Defense Finance receives multiple orders for payments to former spouses in proper form, they are honored in the order received.

Application by Former Spouse

Upon obtaining a final decree dividing a military pension, the decree is forwarded by certified mail to Defense Finance, which has now completed consolidation of the various centers. DD Form 2293, "Request for Former Spouse Payments From Retired Pay," is completed and included with the certified copy of the final decree. If the member was on active duty at the time of dissolution of marriage, the decree must certify that the Soldiers and Sailors Civil Relief Act of 1940, as amended, was complied with, and evidence that the parties were married for 10 years during which time the member was on active duty earning retirement credits must be provided. Recently, Defense Finance has been requesting a copy of the marriage certificate even if the decree recites the 10-year marriage requirement. Husband and wife must be identified with addresses and Social Security numbers. Certification that the decree has not been amended or set aside must be made. A formal military qualified domestic relations order is not required.

Defense Finance is allowed 90 days to respond in writing regarding whether the order will be honored. It is the policy of Defense Finance to honor orders that meet the requirements of the law. Defense Finance will not commence payments prior to the date of service on that office. Defense Finance will hold disputed amounts in a suspense account in the event of an appeal. The member may request reconsideration of any administrative decision reached by Defense Finance concerning honoring court orders.

Family law practitioners familiar with military dissolution cases, and many military retirees, believe that state courts approach the FSPA and equitable distribution statutes by automatically awarding the nonmilitary spouse a mathematical portion of the military retirement utilizing the Deloach formula. If courts do automatically award pension benefits, miscarriages of justice and errors of law are occurring. The following language from the Fifth District Court of Appeal concerning a pension is instructive:

No case from any court has ever held that one spouse must automatically be awarded some portion of the other spouse's pension benefits, irrespective of all other equities and the apportionment of other assets and liabilities. Such a myopic approach to equitable distribution of pension benefits and joint liabilities would conflict not only with *Diffenderfer* and *Bujarski*, but with the admonition in *Canakaris* that the trial court has broad discretionary power to utilize various and interrelated remedies to achieve an equitable result.

Hallman v. Hallman, 575 So. 2d 738, 739 (Fla. 5th DCA 1991). Thus, marital assets may be offset by marital liabilities, or other assets may be awarded to the nonmilitary spouse to achieve equitable distribution. See Johnson v. Johnson, 602 So. 2d 1348 (Fla. 2d DCA 1992). Although the Florida Supreme Court has said that it is generally preferable to treat a military pension as a marital asset, and F.S. §61.076 (1988) says all pensions are marital assets-- they may also be treated as a stream of income from which alimony may be paid. Retirement plans should not be treated simultaneously as both assets and sources of income. The Johnson court also found that a lack of sufficient assets or other circumstances exist which leave the court no choice but to utilize pension benefits in calculating permanent, periodic alimony. Finally, a court can consider lengthy periods of separation of the parties as justification for not including pension benefits in the marital distribution scheme or in choosing a date to value assets. Sheffield v. Sheffield, 522 So 2d 986 (Fla. 1st DCA 1988); Temple v. Temple, 519 So. 2d 1054 (Fla. 4th DCA 1988); Bobb v. Bobb, 552 So. 2d 334 (Fla. 4th DCA 1989). If the military pension is treated as an asset, the factors contained in F.S. §61.075 should be considered and findings of fact made. The parties' contributions to child rearing must be considered, especially where the military member is a "geographic bachelor." Days v. Days, 617 So. 2d 417 (Fla. 1st DCA 1993). However, in a childless marriage where the parties live in separate states for most of a long-term marriage, pursuing their own careers but "dating" on holidays, the pension should not be divided, but rather treated solely as the property of the member in accordance with the FSPA and Florida case law. Whether the nonmilitary spouse should share in "non-passive" post-dissolution increases in pension plans based upon the "foundation of marital effort" theory should be decided by the Florida Supreme Court because of a conflict among circuits. See Boyett v. Boyett, 683 So. 2d 1140 (Fla. 5th DCA 1996).

Whatever the outcome on this important issue, if courts fail to include findings of fact in military pension cases and simply divide the pension using the Deloach formula, *the impression is left that the division is accomplished automatically as a matter of law, instead of equitably as the law requires.* Florida is not a community property state. The general perception of unfairness drives proposals for reform in Congress by both former spouses and military retirees.

Federal Proposals for Reform

Spearheaded by the American Retirees Association and Ex-Partners of Servicemen for Equality, bills seeking to amend FSPA have been introduced on a regular basis since 1982. The suggestions for change have been:

> **H.R. 572** - Terminates the nonmilitary spouse's property interest in military retirement benefits in the event of a remarriage.

> **H.R. 3776** - Creates a presumption that the nonmilitary spouse should receive a pro rata division of retired pay if the couple were married at least 10 years; eliminates 10-year requirements for direct payments.

> **H.R. 2200** - Restricts awards under FSPA to an amount or percentage of the military member's pay at the time of divorce, not retirement; establishes a two-year statute of limitations for former spouses to seek a division of retired pay from time of divorce; reaffirms current prohibition on division of veteran's disability pay.

> **H.R. 3574** - provides former spouse entitlement to separation bonus payout benefits connected to military "right-sizing."

Testimony from the Committee on Veterans' Affairs Hearings:

On September 24, 1997, Congressman Bob Stump (R-AZ) introduced the Uniformed Services Former Spouses' Equity Act of 1997.10 This proposal would have amended the USFSPA in the following particulars:

> Terminate payments of retired pay to a former spouse upon his or her remarriage.

> Base allocation of retired pay on a member's rank and service at the time of the divorce.

> Impose a statute of limitations on efforts to seek a division of retired pay.

> Impose limits on the allocation of disability retired pay to former spouses.

Medical Care and Commissary and Exchange Benefits.

The USFSPA also makes provisions for un-remarried former spouses to receive medical benefits and commissary and exchange privileges. To be eligible, the former spouse must have been married to the member for at least 20 years during a period in which the member performed at least 20 years of military service. However, a former spouse who otherwise qualifies for medical benefits loses coverage if he or she becomes covered by an employer-sponsored health plan. Subsequent changes to the USFSPA allow former spouses who do not meet the 20/20/20 eligibility criterion to receive or participate in reduced or alternative medical benefit plans.

State Legislative Efforts

With the failure of Congress to act, some veterans have made their case to the individual states. To date the most significant legislative initiative is Oklahoma's HR 1053, which is primarily concerned with treating military retainer pay as alimony rather than property. It passed the House and had sufficient votes in the Senate, but has been stalled in committee by family law attorneys afraid of losing their cash cow. For recent developments visit **www.ULSG.com.**

EX-POSE AND DORIS MOZLEY
The Professional Former Spouses

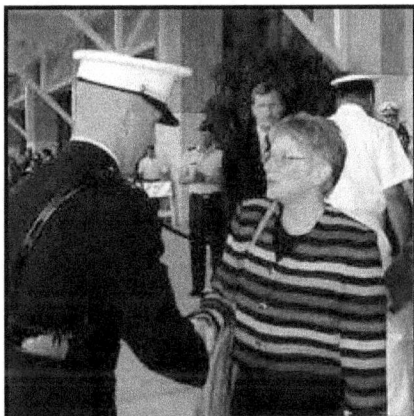

Doris Mozley is a "professional former spouse," i.e. she has made it her life's work to squeeze every dime she can out of her former husband to "punish" him for "dumping" her for another woman, and continues to live as if she herself had served in the Armed Forces. In doing so she has seriously damaged the lives of thousands of veterans who have honorably served our nation – all in the name of her own personal vengeance. Believe it or not she actually wrote, "FSPA was deliberately crafted to give the military member *much more than his fair share*, and that is still the case."

There are a number of "professional former spouses" out there who continually lobby for further division of retainer and disability pay, but the most notorious by far is Doris Mozley.

Ms. Mosley is a former military spouse who was for all intents and purposes a "1950's housewife." She was married for 31 years to her former husband, who was a Navy Officer and a doctor. Sometime after he retired from the U.S. Navy, he told Doris he wanted a divorce.

Doris must have loved the military life and all the trappings that went with being an "officer's wife." The "important" things, like getting your car saluted when you entered the gate at a military installation, and parking closer to the front door of the commissary in the spaces reserved for senior officers. In fact, as recently as 2004, she was a member of the Richmond Area Chapter of the Military Officers Association of America (MOAA). To quote her introductory statement after being introduced by Congressman Bob Stump to the House Committee on Veterans' Affairs:

> "Thank you, Mr. Stump. I thank the Committee on Veterans' Affairs and the Honorable Bob Stump for inviting me to speak here today before this distinguished panel. By way of introduction, I'm Doris Mozley, a 20-year Navy wife, divorced from retired Captain Paul David Mozley, Medical Corps, U.S. Navy Retired, after a 30 year marriage. I have been actively seeking to improve the legal security for military wives since 1980, when I wrote to each member of the Supreme Court, urging it to rule in the McCarty case that military pensions be categorized as community property, and to mandate a pro rata split at divorce. I'm still waiting for a reply. I've been working since that time, now 18 years, in an effort to make it safer to serve our country as a military wife."

This was the introduction to her testimony before Congress on Wednesday, August 5, 1998. What is her self-described objective in life since her divorce? Below are her own words in closing statements before Congress in the same prepared remarks:

> "...And as I said, I've been working in this area 18 years. I've written an awful lot of stuff. It was hard, and I've had a lot of it published in military papers because my objective was to try to influence military

members as well as Congress. I understand that everybody on Capitol Hill reads these papers. I had not even read one when I was a military wife, so I started and we've had some success and I've enclosed a lot of articles from papers. Of course, I don't have the time to go in and you don't have the patience here anyway, but to show you what experts think of what is going on in our country today. The disintegration of family, the neglect of children, the abused. All this deriving from divorce. What we need to do is enact laws that will tend to diminish. It seems so simple."

Doris' idea for enacting laws that tend to diminish military divorce is to require that ALL military members who get divorced from their spouses or vice versa... all military members be required to equally split fifty percent of their regular retirement pay and fifty percent of their VA service connected disability pay as a community property asset. Any child support payments to the former spouse (assuming the former spouse gets custody of any minor children) would be in addition to the community property settlement. She fails to address the issues of fault, financial need, or individual circumstances.

In her world, this would create such a crippling financial disincentive to active duty career and honorably retired military members that they would never consider divorce, and would stay married in large part to avoid the financial penalty. Her logic being that "love of money," and keeping that military retirement paycheck in its entirety, would prevent a retired military member from "dumping" his (her presumptions tend to be that the retired military member is always male) faithful and loyal housewife and mother of his children for a younger and more physically attractive woman.

How narrow minded it is to conclude that the "villain" in every military divorce situation is always the retiring or retired service member, and that it is always a man. For every horror story about a former spouse who was victimized by a retired service member, there is an equally horrifying story about a military member who has been victimized by a former spouse and a biased divorce court judge.

As I said, Ms. Mozley believes that VA disability benefits should also be divisable as community property. Keep in mind, many of the disabilities which the Veterans Administration deems to be service related in nature were received in combat or combat training situations. In addition to wounds sustained in combat, a typical peacetime example would be the back and spinal cord injuries common among paratroopers having a "hard" landing, or those sustained in military specialty which requires heavy lifting of military hardware on a daily basis, i.e. missiles, rockets, bomb racks, pylons, heavy guns and ammo. Then there is hearing loss from hazardous noise, head and neck injuries sustained in wrecks in government vehicles, many types of cancers and diseases caused by exposure to and handling of hazardous materials (i.e. Agent Orange), and the all-too-common Post Tramatic Stress Disorder (PTSD) which arises from being witness of the "horrors of war."

How can anyone with an ounce of conscience consider a service connected disability to be a divisable community property   asset? These service connected disabilities very often negatively impact the quality of the rest of the retiree's life. A former spouse who did *not* get injured and sustain any service connected injuries should not be able to receive half of that financial compensation.

Why, then, is Ms. Mozely pursuing this? Could it be part of a personal vendetta? Remember, her former husband is a retired Navy Captain - that means he retired at the paygrade of 0-6. Doris was eventually awarded fifty percent of his military retirement income as a community property asset. At some point after his retirement, the Veterans Administration, based on medical evidence, determined that Captain Mozley was entitled to ten percent service connected compensation. This ten percent of his regular monthly retirement income was converted under United States Law to VA service connected disability pay. This made the 10% tax exempt *and also exempted it from division as a community property asset.* How did this decision financially impact Doris Mozley's portion of her former husband's 0-6 retirement income? She lost five percent of the income. Instead of getting fifty percent of his retirement pay, because ten percent was deemed VA service connected disability income, Ms. Doris Mozley is now receiving "only" 45% of his retirement income as her community property asset share. And so, in order to recover that 5%, she is willing to push for an Amendment which would cause *all* VA disability compensation to become divisible – including that of combat veterans who have lost limbs and eyes in the service of their country.

Ms. Mosley is completely out of touch with the reality of military marriage and the divorce of young couples today, versus young military couples who married in the 1950s. Today we have "horror stories" of retired *female* officers and NCOs who were physically and verbally abused by their civilian spouses while married. These male former spouses are also entitled under USFSPA to up to half of their honorably military retired ex-wives retirement income. There are documented horror stories of spouses of enlisted military members abandoning their children and their husbands and yet, under USFSPA, these former spouses can also be financially rewarded for their miscreant behavior with up to 50% of the retirement income as a community property asset.

Our active duty military numbers are lower than they were in 1982, and yet the number of Defense Finance Accounting Service (DFAS) monthly court ordered requests for garnishment of military retirement pay as a community property asset is the highest ever. So how can Doris rationalize that stronger amendments to USFSPA biased favor of the former spouse will somehow halt the disintegration of family, and stop the neglect of children? Quite to the contrary, USFSPA has simply created a cottage industry for Family Law attorneys. Win, lose or draw; they always get paid.

You don't even need to be retired to become a USFSPA victim. State Some Divorce Court judges are awarding a portion of military retirement pay to former spouses before they are even retirement eligible! (See the Chapter entitled "Rumsfeld's Town Hall Meeting" for the case of LtCol Patricia Larrabee).

Doris is of course aware that military retirees adversely affected by the USFSPA have become a "band of brothers and sisters," and have filed a class action lawsuit against the United States Government with Donald Rumsfield as the plantiff. Since she has made a career out of being a former military spouse, she is attempting to be actively engaged in the process of putting her spin on the decision making process before the court. Recently she appeared in the courtroom, and afterward sent a letter to the attorney representing the retirees which reveals a lot about how out of touch she is with reality.

In Ms. Mozley's letter to Attorney Katz she refers to the military retirees who are part of the class action lawsuit as "greedy dummies," and states that these military retirees "deserve to be fleeced" and are "selfish beings." She was upset over Attorney Katz behavior in representing his clients, and also stated "... and I do not believe the huge fees you are receiving in your quest to rob (military) widows and orphans justifies rude behavior."

What does USFSPA have to do with military widows and orphans??? The answer: ABSOLUTELY NOTHING! This is a smoke and mirrors technique designed to cloud the real issue. Military orphans have absolutely nothing to do with the Former Spouse Protection Act. Why Doris mentions them is beyond me. The Department of Veterans Affairs provides several financial benefits to military orphans, and several individual U.S. states also provide their own state funded benefits to military orphans. None of them are connected *in any way* to USFSPA.

During the 1st session of the 106th Congress in January of 1999, Arizona Representative Bob Stump introduced House Resolution 72 in an attempt to amend title 10 of the United States Code and revise the rules relating to the court-ordered apportionment of the retired pay of members of the Armed Forces to former spouses. The short Title was the 'Uniformed Services Former Spouses Equity Act of 1999.' This Bill, which had 65 co-sponsors, would have done the following:

TERMINATE PAYMENTS OF RETIREMENT INCOME AS A COMMUNITY PROPERTY ASSET UPON REMARRIAGE OF THE FORMER SPOUSE. Had this become law, a retired military member would not be placed into financial servitude to a former spouse for the rest of his or her natural life. As it stands now, military retirees are the only federal government employees who must continue dividing their retirement pay with a former spouse after that former spouse remarries.

MAKE THE AWARD OF RETIRED PAY BASED ON

THE RETIREE'S LENGTH OF SERVICE AND PAY GRADE AT THE TIME OF DIVORCE. If you divorced as an O-3 or an E-5 your former spouse would not have gotten a financial windfall from any future promotions that you received after the divorce. For example, you get divorced as an E-5, but eventually retire at the E-9 pay grade and are remarried. Why should your first spouse get an even split of a retirement check that they contributed nothing to?

PLACE A LIMITATION ON THE TIME FOR SEEKING DIVISION OF RETIRED PAY. Under existing law, since 1982 there is NO limitation on the time a former spouse may seek division of retired pay. This bill, if passed into law, would have limited the time to within two years of the divorce. All civilian Federal Employees are protected in that there is a one year limit from the time of their divorce for a former spouse to seek part of their retirement income as a community property asset. Only honorably retired veterans are penalized with NO LIMITATION on the time for seeking division of their military retirement pay.

LIMIT APPORTIONMENT OF DISABILITY PAY WHEN RETIRED PAY HAS BEEN WAIVED. The Chairman of the Military Personnel Subcommittee, Steve Buyer (R-IN), could have held hearings on this H.R. 72, but instead chose to ignore it and refused repeated requests for hearings on H.R. 72. Consequently it died in the House without even having a hearing.

What was Doris Mozley's reaction to Rep. Stump and H.R. 72? On May 10, 1999 the ARMY TIMES printed her opinion in their editorial column. It was also posted on the website Protecting Marriage, Inc.

The title "Protecting Marriage, Inc." has a nice "family value" kind sound to its name, doesn't it? However, Protecting Marriage was actually an organization created by former spouses (of which Doris Mozley was a founding member) to educate military former spouses and future former spouses on how to use the legal system to be awarded up to half of a retired service member's retirement income, as well as the legal loop holes required to get their VA service connected disability income as well.

Doris Mozley also chairs the Committee for Justice and Equity for the Military Wife, which is based in Richmond, VA. What exactly is the Committee for Justice and Equity for the Military Wife, and how many members are on this "Military Wife" committee? This is 2004, not 1954 - why isn't this "committee" called the Committee for Justice and Equity for the Military *Spouse*? If Doris' concern is ONLY for women, shouldn't there also be a Committee for Justice and Equity for the Military Retired Woman? There isn't one put together by Ms. Mozley because her intentions are not protecting the rights of women, her intentions are emptying the pockets of retired servicemen of up to half of their retirement pensions and lining the purses of their former wives.

Doris has also helped to create an organization titled EX-POSE, which stands for "Ex-Partners of Servicemembers for Equality." This organization is more obvious in its objectives. EX-POSE promotes the interests of former spouses of all Armed Services personnel, provides information for spouses regarding separation and divorce from active duty or retired military service members, and can inform them about eligibility for a possible share of the military pension and explain the Survivor Benefit Plan (SBP).

This sounds more like a woman hell bent with terminal rage over her own divorce rather than someone that deeply cares about keeping families together for the sake of the children, as she stated in her August 5, 1998 testimony before Congress.

Why create an organization for former spouses that has no real authority? The answers are POLITICS and MONEY. If you can convince a Senator that you have "thousands" of people in your organization with a specific political agenda and you want that Senator's support on the issue, you have to be able to convince that Senator that it is in his or her best interest if they wish to remain elected. Once you establish an organization, if you can get 1,000 to 2,000 people nationwide to send in fifteen dollars in dues, you now have a group organization that has 2,000 members that represent a "grass roots political movement." If you happen to live in Virginia as Doris Mozley does, and you are a politically motivated person with your own divorce agenda, Congressmen and Senators are more likely to take you seriously if you tell them your organization has over 2,000 to 8,000 (Okay... stretch the truth) voting members.

Doris is to be commended for knowing the motivation of career politicians. It is called getting re-elected. Career politicians do not want to lose their jobs. Once they achieve their initial goal of getting into office, the majority of the rest of the time they spend in office is devoted to getting re-elected.

If you would like to have a personal sit down conversation with your elected representatives and tell them your issues, make an annual donation of 2,000 dollars to their office. Money gets their attention much faster than letter writing campaigns. Politicians have staff members that keep track of how much you have donated to their campaign, and you can make "friends" with your political representatives with your money. They might even remember your first name and remember your own personal political issues if you attend enough of their political fund raisers and give generously.

By creating an organization like EX-POSE, and having angry former spouses sending in their annual dues to remain a part of your organization, this creates political "clout" in your behalf. You have money to give to politicians to get their attention, and this gives you the opportunity to do your own personal lobbying and satisfy your own self-serving interests.

Why would you join Doris Mozley's organization just to pay her fifteen dollars for information that is already "free" on the Internet? She has been receiving 45 Percent of the retirement check of a retired 0-6 PLUS yearly cost of living allowances (COLAS). She has her own military dependent I.D. card even though they have been divorced for over 24 years and she also can use the BX, Commissary, and Tricare. Even though she has been divorced from her EX for over 24 years now, I'd be willing to wager that she still parks in the 'Reserved for Colonel' Parking space at her local commissary.

Social Security Laws state that if she has not remarried, she is entitled to her former spouses' social security as well. Her former spouse was a Doctor. That sounds like a pretty hefty social security check. Coupled with the retirement money she receives from his military career, I would be willing to bet she isn't homeless, unlike many veterans. Your fifteen dollars does not have to be accounted for. Where do you think the money is going?

A bullet statement from the literature EX-POSE sent to me...

Remember her introduction to her testimony before Congress on Wednesday, August 5, 1998...

What's Mine is Mine, What's Yours is Mine

"...And as I said, I've been working in this area 18 years. I've written an awful lot of stuff... The disintegration of family, the neglect of children, the abused. All this deriving from divorce. What we need to do is enact laws that will tend to diminish. It seems so simple."

Ms. Mozley speaks like a true politician. She declares in words before Congress that her goals (so to speak) are noble and about keeping marriages together, yet at the same time she offers plenty of information to prospective former spouses of active duty and retired service members on how to go after the retirement money as a community property asset. She is using language to deceive, and is concealing and misrepresenting her true goals for military wives. That is called "doublespeak." A perfect example of that is an article she wrote entitled *Fairness' Amendment Makes Little Sense for Former Spouses.* Here are a few excerpts, along with relavent commentary:

"With the passage of the weak former spouses act in 1982, ex-military spouses received some "justness," but the bill was far from impartial because it was designed to give a spouse minimal protection and was hedged with qualifications designed to make it difficult for her to receive her share. A fair share was an absolute impossibility."

"FSPA was deliberately crafted to give the military member much more than his fair share, and that is still the case."

"One of the provisions of Stump's bill is to limit payments even for child support. While other fathers in our country are going to jail if they refuse to pay child support, military members want to be excused from so onerous a burden as putting bread into their own children's mouths."

"Another Stump proposal is to revoke pension shares awarded as property upon the remarriage of former spouses. And to keep any ex-spouse from escaping, this would apply to all spouse remarriages, even those entered into before Stump's "fairness" amendment

became law."

"Ninety-five percent of divorced service members remarry during that first year. Apparently, Stump thinks it's OK for a service member to have a normal, married life, and not be penalized, but not their former spouse. In fact, a service member is rewarded because his second - or maybe third - spouse is instantly entitled to an array of taxpayer benefits solely on the basis of a marriage certificate to a retired military member."

On the other hand, a wife loses all their perks except their pension share if they remarry. Talk about discrimination! The ABA testified that such a scheme would lead to a large amount of litigation without any corresponding improvement in achieving equitable distribution of property and no apparent federal policy would be served by such a pre-emption of state law.

The end result would be to increase the number of military spouses who wouldn't share in the benefits that their many years of work helped produce.

This amendment would only cater to the greed of divorced military members. Mr. Stump's "statute of limitations" is a misnomer.

Attorney Marshal S. Willick, testifying for the ABA, said, "It is difficult to conceive of any set of facts that would make it equitable for one souse to continue receiving the other spouse's rightful share of the property in the future, but that is exactly what the current proposal would allow."

"Mr. Stump's plan to compute a spouse's share as of the date of divorce would enable a service member to continue serving without compensating the spouse in any way for delay of receipt of their share."

"In sum, if we want children to be denied bread to eat; a spouse to be penalized for entering into an honorable marriage - if we want to help a member steal from their former spouse, and if we want a spouse to be cheated out of a true pro rata share of the pension they helped to earn, Congress should accept Stump's Orwellian definition of 'equity' and pass his despicable amendments designed only to help service members evade duty to their spouses, children and society."

THE ARA

Fighting the Good Fight

American Retirees Association

On Alert for Military Service Members

"The Price of Liberty is Eternal Vigilance"

A leading proponent of reform has been the American Retirees Association (ARA), a California-based nonprofit organization of active duty, reserve and retired members, both male and female. The ARA was formed in 1984 for the exclusive purpose of curing the inequities in the USFSPA.

This single-minded objective is pursued closely with the other military veterans' organizations whose broader agendas include USFSPA reform or repeal. These groups agree that the following would go far toward restoring some fairness and equity to the military divorce process:

Terminate USFSPA payments upon the remarriage of the benefiting ex-spouse, while ending current payments to remarried former spouses not more than 180 days from the date of enactment of the amendment. This provision could be softened by providing that it would not apply to ex-spouses who had remarried but were subsequently widowed or divorce and are currently unmarried.

Restrict awards under the USFSPA to correspond to the military retiree's length of service and pay grade at the time of divorce, not at the time of retirement. USFSPA payments would, however, be adjusted to existing pay scales at the time of retirement.

Establish a statute of limitations giving former spouses two years from the date of a final divorce to seek a division of military retirement pay under the USFSPA.

Reinforce the provision of the USFSPA that precludes the inclusion of disability pay in the calculation of disposable pay. Consider, however, including disability pay wherever military retired pay is the only asset of marriage.

Provide specific wording to protect active-duty military personnel by precluding: involuntary, court-ordered retirements in order to commence USFSPA payments; distribution of active-duty pay pursuant to court orders under the USFSPA; and USFSPA payments after recall of retired military members to active duty.

Preclude retroactive application of the USFSPA for any divorce finalized prior to 1 February 1983. Public Law 101-510 of 5 November 1990 prohibits retroactive opening of divorces finalized on and before 25 June 1981 - one day before the U.S. Supreme Court's McCarty decision. This denies relief for those divorced during the gap between McCarty and the effective date of the USFSPA. The failure to grandfather the USFSPA was a manifest injustice to military people who had served honorably prior to 1 February 1983 and were peremptorily deprived of their matured right to full retirement pay. Let's require the leadership of the uniformed services to brief their personnel on the *existence* and *significance* of USFSPA!

THE LAW

10 U.S.C. § 1408
Sec. 1408. Payment of Retired or Retainer Pay in Compliance With Court Orders

(a) Definitions. - In this section:

(1) The term "court" means –

(A) any court of competent jurisdiction of any State, the District of Columbia, the Commonwealth of Puerto Rico, Guam, American Samoa, the Virgin Islands, the Northern Mariana Islands, and the Trust Territory of the Pacific Islands;

(B) any court of the United States (as defined in section 451 of title 28) having competent jurisdiction; and

(C) any court of competent jurisdiction of a foreign country with which the United States has an agreement requiring theUnited States to honor any court order of such country.

(2) The term "court order" means a final decree of divorce, dissolution, annulment, or legal separation issued by a court, or a court ordered, ratified, or approved property settlement incident to such a decree (including a final decree modifying the terms of a previously issued decree of divorce, dissolution, annulment, or legal separation, or a court ordered, ratified, or approved property settlement incident to such previously issued decree), which –

(A) is issued in accordance with the laws of the jurisdiction of that court;

(B) provides for –

(i) payment of child support (as defined in section 462(b) of the Social Security Act (42 U.S.C. 662(b)));

(ii) payment of alimony (as defined in section 462(c) of the Social Security Act (42 U.S.C. 662(c))); or

(iii) division of property (including a division of community property); and

(C) in the case of a division of property, specifically provides for the payment of an amount, expressed in dollars or as a percentage of disposable retired pay, from the disposable retired pay of a member to the spouse or former spouse of that member.

(3) The term "final decree" means a decree from which no appeal may be taken or from which no appeal has been taken with in the time allowed for taking such appeals under the laws applicable to such appeals, or a decree from which timely appeal has been taken and such appeal has been finally decided under the laws applicable to such appeals.

(4) The term "disposable retired pay" means the total monthly retired pay to which a member is entitled less amounts which –

(A) are owed by that member to the United States for previous overpayments of retired pay and for recoupments required by law resulting from entitlement to retired pay;

(B) are deducted from the retired pay of such member as a result of forfeitures of retired pay ordered by a court-martial or as a result of a waiver of retired pay required by law in order to receive compensation under title 5 or title 38;

(C) in the case of a member entitled to retired pay under chapter 61 of this title, are equal to the amount of retired pay of the member under that chapter computed using the percentage of the member's disability on the date when the member was retired (or the date on which the member's name was placed on the temporary disability retired list); or

(D) are deducted because of an election under chapter 73 of this title to provide an annuity to a spouse or former spouse to whom payment of a portion of such member's retired pay is being made pursuant to a court order under this section.

(5) The term "member" includes a former member entitled to retired pay under section 1331 (FOOTNOTE 1) of this title.

(FOOTNOTE 1) See References in Text note below.

(6) The term "spouse or former spouse" means the husband or wife, or former husband or wife, respectively, of a member who, on or before the date of a court order, was married to that member.

(7) The term "retired pay" includes retainer pay.

(b) Effective Service of Process. - For the purposes of this section –

(1) service of a court order is effective if –

(A) an appropriate agent of the Secretary concerned designated for receipt of service of court orders under regulations prescribed pursuant to subsection (i) or, if no agent has been so designated, the Secretary concerned, is personally served or is served by certified or registered mail, return receipt requested;

(B) the court order is regular on its face;

(C) the court order or other documents served with the court order identify the member concerned and include, if possible, the social security number of such member; and

(D) the court order or other documents served with the court order certify that the rights of the member under the Soldiers' and Sailors' Civil Relief Act of 1940 (50 U.S.C. App. 501 et seq.) were observed; and

(2) a court order is regular on its face if the order –

(A) is issued by a court of competent jurisdiction;

(B) is legal in form; and

(C) includes nothing on its face that provides reasonable

notice that it is issued without authority of law.

(c) Authority for Court To Treat Retired Pay as Property of the Member and Spouse.

(1) Subject to the limitations of this section, a court may treat disposable retired pay payable to a member for pay periods beginning after June 25, 1981, either as property solely of the member or as property of the member and his spouse in accordance with the law of the jurisdiction of such court. A court may not treat retired pay as property in any proceeding to divide or partition any amount of retired pay of a member as the property of the member and the member's spouse or former spouse if a final decree of divorce, dissolution, annulment, or legal separation (including a court ordered, ratified, or approved property settlement incident to such decree) affecting the member and the member's spouse or former spouse

(A) was issued before June 25, 1981, and

(B) did not treat (or reserve jurisdiction to treat) any amount of retired pay of the member as property of the member and the member's spouse or former spouse.

(2) Notwithstanding any other provision of law, this section does not create any right, title, or interest which can be sold, assigned, transferred, or otherwise disposed of (including by inheritance) by a spouse or former spouse. Payments by the Secretary concerned under subsection

(d) to a spouse or former spouse with respect to a division of retired pay as the property of a member and the member's spouse under this subsection may not be treated as amounts received as retired pay for service in the uniformed services.

(3) This section does not authorize any court to order a member to apply for retirement or retire at a particular time in order to effectuate any payment under this section.

(4) A court may not treat the disposable retired pay of a

member in the manner described in paragraph

(1) unless the court has jurisdiction over the member by reason of -

(A) his residence, other than because of military assignment, in the territorial jurisdiction of the court,

(B) his domicile in the territorial jurisdiction of the court, or

(C) his consent to the jurisdiction of the court.

(d) Payments by Secretary Concerned To Spouse or Former Spouse. –

(1) After effective service on the Secretary concerned of a court order providing for the payment of child support or alimony or, with respect to a division of property, specifically providing for the payment of an amount of the disposable retired pay from a member to the spouse or a former spouse of the member, the Secretary shall make payments (subject to the limitations of this section) from the disposable retired pay of the member to the spouse or former spouse in an amount sufficient to satisfy the amount of child support and alimony set forth in the court order and, with respect to a division of property, in the amount of disposable retired pay specifically provided for in the court order. In the case of a member entitled to receive retired pay on the date of the effective service of the court order, such payments shall begin not later than 90 days after the date of effective service. In the case of a member not entitled to receive retired pay on the date of the effective service of the court order, such payments shall begin not later than 90 days after the date on which the member first becomes entitled to receive retired pay.

(2) If the spouse or former spouse to whom payments are to be made under this section was not married to the member for a period of 10 years or more during which the member

performed at least 10 years of service creditable in determining the member's eligibility for retired pay, payments may not be made under this section to the extent that they include an amount resulting from the treatment by the court under subsection -

(c) of disposable retired pay of the member as property of the member or property of the member and his spouse.

(3) Payments under this section shall not be made more frequently than once each month, and the Secretary concerned shall not be required to vary normal pay and disbursement cycles for retired pay in order to comply with a court order.

(4) Payments from the disposable retired pay of a member pursuant to this section shall terminate in accordance with the terms of the applicable court order, but not later than the date of the death of the member or the date of the death of the spouse or former spouse to whom payments are being made, whichever occurs first.

(5) If a court order described in paragraph (1) provides for a division of property (including a division of community property) in addition to an amount of child support or alimony or the payment of an amount of disposable retired pay as the result of the court's treatment of such pay under subsection (c) as property of the member and his spouse, the Secretary concerned shall pay (subject to the limitations of this section) from the disposable retired pay of the member to the spouse or former spouse of the member, any part of the amount payable to the spouse or former spouse under the division of property upon effective service of a final court order of garnishment of such amount from such retired pay.

(e) Limitations. –

(1) The total amount of the disposable retired pay of a member payable under all court orders pursuant to

subsection (c) may not exceed 50 percent of such disposable retired pay.

(2) In the event of effective service of more than one court order which provide for payment to a spouse and one or more former spouses or to more than one former spouse, the disposable retired pay of the member shall be used to satisfy (subject to the limitations of paragraph (1)) such court orders on a first-come, first-served basis. Such court orders shall be satisfied (subject to the limitations of paragraph (1)) out of that amount of disposable retired pay which remains after the satisfaction of all court orders which have been previously served.

(3)(A) In the event of effective service of conflicting court orders under this section which assert to direct that different amounts be paid during a month to the same spouse or former spouse of the same member, the Secretary concerned shall –

(i) pay to that spouse from the member's disposable retired pay the least amount directed to be paid during that month by any such conflicting court order, but not more than the amount of disposable retired pay which remains available for payment of such court orders based on when such court orders were effectively served and the limitations of paragraph (1) and subparagraph (B) of paragraph (4);

(ii) retain an amount of disposable retired pay that is equal to the lesser of –

(I) the difference between the largest amount required by any conflicting court order to be paid to the spouse or former spouse and the amount payable to the spouse or former spouse under clause (i); and –

(II) the amount of disposable retired pay which remains available for payment of any conflicting court order based on when such court order was effectively served and the limitations of paragraph (1) and subparagraph (B) of paragraph (4); and –

(iii) pay to that member the amount which is equal to the amount of that member's disposable retired pay (less any amount paid during such month pursuant to legal process served under section 459 of the Social Security Act (42 U.S.C. 659) and any amount paid during such month pursuant to court orders effectively served under this section, other than such conflicting court orders) minus –

(I) the amount of disposable retired pay paid under clause (i); and –

(II) the amount of disposable retired pay retained under clause (ii).

(B) The Secretary concerned shall hold the amount retained under clause (ii) of subparagraph (A) until such time as that Secretary is provided with a court order which has been certified by the member and the spouse or former spouse to be valid and applicable to the retained amount. Upon being provided with such an order, the Secretary shall pay the retained amount in accordance with the order.

(4)(A) In the event of effective service of a court order under this section and the service of legal process pursuant to section 459 of the Social Security Act (42 U.S.C. 659), both of which provide for payments during a month from the same member, satisfaction of such court orders and legal process from the retired pay of the member shall be on a first-come, first-served basis. Such court orders and legal process shall be satisfied out of moneys which are subject to such orders and legal process and which remain available in accordance with the limitations of paragraph (1) and subparagraph (B) of this paragraph during such month after the satisfaction of all court orders or legal process which have been previously served.

(B) Notwithstanding any other provision of law, the total amount of the disposable retired pay of a member payable by the Secretary concerned under all court orders pursuant to this section and all legal processes pursuant to section 459 of the Social Security Act (42 U.S.C. 659) with respect to a member may not exceed 65 percent of the amount of the retired pay payable to such member that is considered under section 462 of the Social Security Act (42 U.S.C. 662) to be remuneration for employment that is payable by the United States.

(5) A court order which itself or because of previously served court orders provides for the payment of an amount which exceeds the amount of disposable retired pay available for payment because of the limit set forth in paragraph (1), or which, because of previously served court orders or legal process previously served under section 459 of the Social Security Act (42 U.S.C. 659), provides for payment of an amount that exceeds the maximum amount permitted under paragraph (1) or subparagraph (B) of paragraph (4), shall not be considered to be irregular on its face solely for that reason. However, such order shall be considered to be fully satisfied for purposes of this section by the payment to the spouse or former spouse of the maximum amount of disposable retired pay permitted under paragraph (1) and subparagraph (B) of paragraph (4).

(6) Nothing in this section shall be construed to relieve a member of liability for the payment of alimony, child support, or other payments required by a court order on the grounds that payments made out of disposable retired pay under this section have been made in the maximum amount permitted under paragraph (1) or subparagraph (B) of paragraph (4). Any such unsatisfied obligation of a member may be enforced by any means available under law other

than the means provided under this section in any case in which the maximum amount permitted under paragraph (1) has been paid and under section 459 of the Social Security Act (42 U.S.C. 659) in any case in which the maximum amount permitted under subparagraph (B) of paragraph (4) has been paid.

(f) Immunity of Officers and Employees of United States.

(1) The United States and any officer or employee of the United States shall not be liable with respect to any payment made from retired pay to any member, spouse, or former spouse pursuant to a court order that is regular on its face if such payment is made in accordance with this section and the regulations prescribed pursuant to subsection (i).

(2) An officer or employee of the United States who, under regulations prescribed pursuant to subsection (i), has the duty to respond to interrogatories shall not be subject under any law to any disciplinary action or civil or criminal liability or penalty for, or because of, any disclosure of information made by him in carrying out any of his duties which directly or indirectly pertain to answering such interrogatories.

(g) Notice To Member of Service of Court Order On Secretary Concerned. - A person receiving effective service of a court order under this section shall, as soon as possible, but not later than 30 days after the date on which effective service is made, send a written notice of such court order (together with a copy of such order) to the member affected by the court order at his last known address.

(h) Benefits for Dependents Who Are Victims of Abuse by Members Losing Right to Retired Pay. –

(1) If, in the case of a member or former member of the armed forces referred to in paragraph (2)(A), a court order provides (in the manner applicable to a division of property)

for the payment of an amount from the disposable retired pay of that member or former member (as certified under paragraph (4)) to an eligible spouse or former spouse of that member or former member, the Secretary concerned, beginning upon effective service of such court order, shall pay that amount in accordance with this subsection to such spouse or former spouse.

(2) A spouse or former spouse of a member or former member of the armed forces is eligible to receive payment under this subsection if –

(A) the member or former member, while a member of the armed forces and after becoming eligible to be retired from the armed forces on the basis of years of service, has eligibility to receive retired pay terminated as a result of misconduct while a member involving abuse of a spouse or dependent child (as defined in regulations prescribed by the Secretary of Defense or, for the Coast Guard when it is not operating as a service in the Navy, by the Secretary of Transportation); and

(B) the spouse or former spouse –

(i) was the victim of the abuse and was married to the member or former member at the time of that abuse; or

(ii) is a natural or adopted parent of a dependent child of the member or former member who was the victim of the abuse.

(3) The amount certified by the Secretary concerned under paragraph (4) with respect to a member or former member of the armed forces referred to in paragraph (2)(A) shall be deemed to be the disposable retired pay of that member or former member for the purposes of this subsection.

(4) Upon the request of a court or an eligible spouse or former spouse of a member or former member of the armed forces referred to in paragraph (2)(A) in connection with a

civil action for the issuance of a court order in the case of that member or former member, the Secretary concerned shall determine and certify the amount of the monthly retired pay that the member or former member would have been entitled to receive as of the date of the certification –

(A) if the member or former member's eligibility for retired pay had not been terminated as described in paragraph (2)(A); and

(B) if, in the case of a member or former member not in receipt of retired pay immediately before that termination of eligibility for retired pay, the member or former member had retired on the effective date of that termination of eligibility.

(5) A court order under this subsection may provide that whenever retired pay is increased under section 1401a of this title (or any other provision of law), the amount payable under the court order to the spouse or former spouse of a member or former member described in paragraph (2)(A) shall be increased at the same time by the percent by which the retired pay of the member or former member would have been increased if the member or former member were receiving retired pay.

(6) Notwithstanding any other provision of law, a member or former member of the armed forces referred to in paragraph (2)(A) shall have no ownership interest in, or claim against, any amount payable under this section to a spouse or former spouse of the member or former member.

(7)(A) If a former spouse receiving payments under this subsection with respect to a member or former member referred to in paragraph (2)(A) marries again after such payments begin, the eligibility of the former spouse to receive further payments under this subsection shall terminate on the date of such marriage.

(B) A person's eligibility to receive payments under this

subsection that is terminated under subparagraph (A) by reason of remarriage shall be resumed in the event of the termination of that marriage by the death of that person's spouse or by annulment or divorce. The resumption of payments shall begin as of the first day of the month in which that marriage is so terminated. The monthly amount of the payments shall be the amount that would have been paid if the continuity of the payments had not been interrupted by the marriage.

(8) Payments in accordance with this subsection shall be made out of funds in the Department of Defense Military Retirement Fund established by section 1461 of this title or, in the case of the Coast Guard, out of funds appropriated to the Department of Transportation for payment of retired pay for the Coast Guard.

(9)(A) A spouse or former spouse of a member or former member of the armed forces referred to in paragraph (2)(A), while receiving payments in accordance with this subsection, shall be entitled to receive medical and dental care, to use commissary and exchange stores, and to receive any other benefit that a spouse or a former spouse of a retired member of the armed forces is entitled to receive on the basis of being a spouse or former spouse, as the case may be, of a retired member of the armed forces in the same manner as if the member or former member referred to in paragraph (2)(A) was entitled to retired pay.

(B) A dependent child of a member or former member referred to in paragraph (2)(A) who was a member of the household of the member or former member at the time of the misconduct described in paragraph (2)(A) shall be entitled to receive medical and dental care, to use commissary and exchange stores, and to have other benefits provided to dependents of retired members of the armed forces in the same manner as if the member or former member referred to in paragraph (2)(A) was entitled to retired pay.

(C) If a spouse or former spouse or a dependent child eligible or entitled to receive a particular benefit under this paragraph is eligible or entitled to receive that benefit under another provision of law, the eligibility or entitlement of that spouse or former spouse or dependent child to such benefit shall be determined under such other provision of law instead of this paragraph.

(10)(A) For purposes of this subsection, in the case of a member of the armed forces who has been sentenced by a court-martial to receive a punishment that will terminate the eligibility of that member to receive retired pay if executed, the eligibility of that member to receive retired pay may, as determined by the Secretary concerned, be considered terminated effective upon the approval of that sentence by the person acting under section 860(c) of this title (article 60(c) of the Uniform Code of Military Justice).

(B) If each form of the punishment that would result in the termination of eligibility to receive retired pay is later remitted, set aside, or mitigated to a punishment that does not result in the termination of that eligibility, a payment of benefits to the eligible recipient under this subsection that is based on the punishment so vacated, set aside, or mitigated shall cease. The cessation of payments shall be effective as of the first day of the first month following the month in which the Secretary concerned notifies the recipient of such benefits in writing that payment of the benefits will cease. The recipient may not be required to repay the benefits received before that effective date (except to the extent necessary to recoup any amount that was erroneous when paid).

(11) In this subsection, the term "dependent child", with respect to a member or former member of the armed forces referred to in paragraph (2)(A), means an unmarried legitimate child, including an adopted child or a stepchild of the member or former member, who –

(A) is under 18 years of age;

(B) is incapable of self-support because of a mental or physical incapacity that existed before becoming 18 years of age and is dependent on the member or former member for over one-half of the child's support; or

(C) if enrolled in a full-time course of study in an institution of higher education recognized by the Secretary of Defense for the purposes of this subparagraph, is under 23 years of age and is dependent on the member or former member for over one-half of the child's support.

(i) Regulations. - The Secretaries concerned shall prescribe uniform regulations for the administration of this section. (Added Pub. L. 97-252, title X, Sec. 1002(a), Sept. 8, 1982, 96 Stat. 730; amended Pub. L. 98-525, title VI, Sec. 643(a)-(d), Oct. 19, 1984, 98 Stat. 2547; Pub. L. 99-661, div. A, title VI, Sec. 644(a), Nov. 14, 1986, 100 Stat. 3887; Pub. L. 100-26, Sec. 3(3), 7(h)(1), Apr. 21, 1987, 101 Stat. 273, 282; Pub. L. 101-189, div. A, title VI, Sec. 653(a)(5), title XVI, Sec. 1622(e)(6), Nov. 29, 1989, 103 Stat. 1462, 1605; Pub. L. 101-510, div. A, title V, Sec. 555(a)-(d), (f), (g), Nov. 5, 1990, 104 Stat. 1569, 1570; Pub. L. 102-190, div. A, title X, Sec. 1061(a)(7), Dec. 5, 1991, 105 Stat. 1472; Pub. L. 102-484, div. A, title VI, Sec. 653(a), Oct. 23, 1992, 106 Stat. 2426; Pub. L. 103-160, div. A, title V, Sec. 555(a), (b), title XI, Sec. 1182(a)(2), Nov. 30, 1993, 107 Stat. 1666, 1771.)

REFERENCES IN TEXT

Section 1331 of this title, referred to in subsec. (a)(5), was renumbered section 12731 of this title and amended generally by Pub. L. 103-337, div. A, title XVI, Sec. 1662(j)(1), Oct. 5, 1994, 108 Stat. 2998, 2999. A new section 1331 was added by section 1662(j)(7) of Pub. L. 103-337.

The Soldiers' and Sailors' Civil Relief Act, referred to in subsec. (b)(1)(D), is act Oct. 17, 1940, ch. 888, 54 Stat. 1178, as amended, which is classified to section 501 et seq. of the Appendix to Title 50, War and National Defense. For complete classification of this Act to the Code, see section 501 of the Appendix to Title 50 and Tables.

AMENDMENTS

1993 - Subsecs. (b)(1)(A), (f)(1), (2). Pub. L. 103-160, Sec. 1182(a)(2)(A), substituted "subsection (i)" for "subsection (h)". Subsec. (h)(2)(A). Pub. L. 103-160, Sec. 555(b)(1), inserted "or, for the Coast Guard when it is not operating as a service in the Navy, by the Secretary of Transportation" after "Secretary of Defense".
Subsec. (h)(4)(B). Pub. L. 103-160, Sec. 1182(a)(2)(B), inserted "of" after "of that termination".
Subsec. (h)(8). Pub. L. 103-160, Sec. 555(b)(2), inserted before period at end "or, in the case of the Coast Guard, out of funds appropriated to the Department of Transportation for payment of retired pay for the Coast Guard".
Subsec. (h)(10), (11). Pub. L. 103-160, Sec. 555(a), added par. (10) and redesignated former par. (10) as (11).
1992 - Subsecs. (h), (i). Pub. L. 102-484 added subsec. (h) and redesignated former subsec. (h) as (i).
1991 - Pub. L. 102-190 inserted "or retainer" after "retired" in section catch line.
1990 - Pub. L. 101-510, Sec. 555(f)(2), substituted "retired

pay" for "retired or retainer pay" in section catch line.
Subsec. (a). Pub. L. 101-510, Sec. 555(g)(1), inserted heading.
Subsec. (a)(2)(C). Pub. L. 101-510, Sec. 555(f)(2), substituted "retired pay" for "retired or retainer pay" wherever appearing.
Subsec. (a)(4). Pub. L. 101-510, Sec. 555(f)(2), substituted "retired pay" for "retired or retainer pay" wherever appearing in introductory provisions and in subpar. (D).
Subsec. (a)(4)(A). Pub. L. 101-510, Sec. 555(b)(1), inserted before semicolon at end "for previous overpayments of retired pay and for recoupments required by law resulting from entitlement to retired pay".
Subsec. (a)(4)(B). Pub. L. 101-510, Sec. 555(b)(2), added subpar.
(B) and struck out former subpar. (B) which read as follows: "are required by law to be and are deducted from the retired or retainer pay of such member, including fines and forfeitures ordered by courts-martial, Federal employment taxes, and amounts waived in order to receive compensation under title 5 or title 38;".
Subsec. (a)(4)(C) to (F). Pub. L. 101-510, Sec. 555(b)(3), (4), redesignated subpars. (E) and (F) as (C) and (D), respectively, and struck out former subpars. (C) and (D) which read as follows:
"(C) are properly withheld for Federal, State, or local income tax purposes, if the withholding of such amounts is authorized or required by law and to the extent such amounts withheld are not greater than would be authorized if such member claimed all dependents to which he was entitled;
"(D) are withheld under section 3402(i) of the Internal Revenue Code of 1986 if such member presents evidence of a tax obligation which supports such withholding;".

Subsec. (a)(7). Pub. L. 101-510, Sec. 555(f)(1), added par. (7).
Subsec. (b). Pub. L. 101-510, Sec. 555(g)(2), inserted heading.
Subsec. (c). Pub. L. 101-510, Sec. 555(g)(3), inserted heading.
Subsec. (c)(1). Pub. L. 101-510, Sec. 555(f)(2), substituted "retired pay" for "retired or retainer pay".
Pub. L. 101-510, Sec. 555(a), inserted at end "A court may not treat retired pay as property in any proceeding to divide or partition any amount of retired pay of a member as the property of the member and the member's spouse or former spouse if a final decree of divorce, dissolution, annulment, or legal separation (including a court ordered, ratified, or approved property settlement incident to such decree) affecting the member and the member's spouse or former spouse (A) was issued before June 25, 1981, and (B) did not treat (or reserve jurisdiction to treat) any amount of retired pay of the member as property of the member and the member's spouse or former spouse."
Subsec. (c)(2). Pub. L. 101-510, Sec. 555(c), inserted at end "Payments by the Secretary concerned under subsection (d) to a spouse or former spouse with respect to a division of retired pay as the property of a member and the member's spouse under this subsection may not be treated as amounts received as retired pay for service in the uniformed services."
Subsec. (c)(4). Pub. L. 101-510, Sec. 555(f)(2), substituted "retired pay" for "retired or retainer pay".
Subsec. (d). Pub. L. 101-510, Sec. 555(g)(4), inserted heading.
Pub. L. 101-510, Sec. 555(f)(2), substituted "retired pay" for "retired or retainer pay" wherever appearing.
Subsec. (e). Pub. L. 101-510, Sec. 555(g)(5), inserted

heading.

Pub. L. 101-510, Sec. 555(f)(2), substituted "retired pay" for "retired or retainer pay" wherever appearing.

Subsec. (e)(1). Pub. L. 101-510, Sec. 555(d)(1), substituted "payable under all court orders pursuant to subsection (c)" for "payable under subsection (d)".

Subsec. (e)(4)(B). Pub. L. 101-510, Sec. 555(d)(2), substituted "the amount of the retired pay payable to such member that is considered under section 462 of the Social Security Act (42 U.S.C. 662) to be remuneration for employment that is payable by the United States" for "the disposable retired or retainer pay payable to such member".

Subsec. (f). Pub. L. 101-510, Sec. 555(g)(6), inserted heading.

Subsec. (f)(1). Pub. L. 101-510, Sec. 555(f)(2), substituted "retired pay" for "retired or retainer pay".

Subsec. (g). Pub. L. 101-510, Sec. 555(g)(7), inserted heading.

Subsec. (h). Pub. L. 101-510, Sec. 555(g)(8), inserted heading.

1989 - Subsec. (a)(1), (2). Pub. L. 101-189, Sec. 1622(e)(6), substituted "The term 'court" for " 'Court" in introductory provisions.

Subsec. (a)(3). Pub. L. 101-189, Sec. 1622(e)(6), substituted "The term 'final" for " 'Final".

Subsec. (a)(4). Pub. L. 101-189, Sec. 1622(e)(6), substituted "The term 'disposable" for " 'Disposable" in introductory provisions.

Subsec. (a)(4)(D). Pub. L. 101-189, Sec. 653(a)(5)(A), struck out "(26 U.S.C. 3402(i))" after "Code of 1986".

Subsec. (a)(5). Pub. L. 101-189, Sec. 653(a)(5)(B), 1622(e)(6), substituted "The term 'member" for " 'Member" and inserted "entitled to retired pay under section 1331 of this title" after "a former member".

Subsec. (a)(6). Pub. L. 101-189, Sec. 1622(e)(6), substituted "The term 'spouse" for " 'Spouse".

1987 - Subsec. (a)(4). Pub. L. 100-26, Sec. 3(3), made technical amendment to directory language of Pub. L. 99-661, Sec. 644(a). See 1986 Amendment note below.

Subsec. (a)(4)(D). Pub. L. 100-26, Sec. 7(h)(1), substituted "Internal Revenue Code of 1986" for "Internal Revenue Code of 1954".

1986 - Subsec. (a)(4). Pub. L. 99-661, Sec. 644(a), as amended by Pub. L. 100-26, Sec. 3(3), struck out "(other than the retired pay of a member retired for disability under chapter 61 of this title)" before "less amounts" in introductory text, added subpar. (E), and struck out former subpar. (E) which read as follows: "are deducted as Government life insurance premiums (not including amounts deducted for supplemental coverage); or".

1984 - Subsec. (a)(2)(C). Pub. L. 98-525, Sec. 643(a), inserted "in the case of a division of property,".

Subsec. (b)(1)(C). Pub. L. 98-525, Sec. 643(b), inserted ", if possible,".

Subsec. (d)(1). Pub. L. 98-525, Sec. 643(c)(1), substituted "After effective service on the Secretary concerned of a court order providing for the payment of child support or alimony or, with respect to a division of property, specifically providing for the payment of an amount of the disposable retired or retainer pay from a member to the spouse or a former spouse of the member, the Secretary shall make payments (subject to the limitations of this section) from the disposable retired or retainer pay of the member to the spouse or former spouse in an amount sufficient to satisfy the amount of child support and alimony set forth in the court order and, with respect to a division of property, in the amount of disposable retired or retainer pay specifically provided for in the court order" for "After effective service on the Secretary concerned of a court order with respect to the payment of a portion of the retired or retainer pay of a member to the spouse or a former spouse of the member, the Secretary shall, subject to the limitations of this section, make payments to the spouse or former spouse in the amount of the disposable retired or retainer pay of the member specifically provided for in the court order".

Subsec. (d)(5). Pub. L. 98-525, Sec. 643(c)(2), substituted "child support or alimony or the payment of an amount of disposable retired or retainer pay as the result of the court's treatment of such pay under subsection (c) as property of the member and his spouse, the Secretary concerned shall pay (subject to the limitations of this section) from the disposable retired or retainer pay of the member to the spouse or former spouse of the member, any part" for "disposable retired or retainer pay, theSecretary concerned shall, subject to the limitations of this section, pay to the spouse or former spouse of the member, from the disposable retired or retainer pay of the member, any part".

Subsec. (e)(2). Pub. L. 98-525, Sec. 643(d)(1), substituted ", the disposable retired or retainer pay of the member" for "from the disposable retired or retainer pay of a member, such pay" before "shall be used to satisfy".

Subsec. (e)(3)(A). Pub. L. 98-525, Sec. 643(d)(2)(A), struck out "from the disposable retired or retainer pay" before "of the same member".

Subsec. (e)(3)(A)(i). Pub. L. 98-525, Sec. 643(d)(2)(B), substituted "from the member's disposable retired or retainer pay the least amount" for "the least amount of disposable retired or retainer pay" before "directed to be paid".

Subsec. (e)(2)(A)(ii)(I). Pub. L. 98-525, Sec. 643(d)(2)(C), struck out "of retired or retainer pay" before "required by any conflicting".

Subsec. (e)(4)(A). Pub. L. 98-525, Sec. 643(d)(3), struck out "the retired or retainer pay of" before "the same member" and substituted "satisfaction of such court orders and legal process from the retired or retainer pay of the members shall be" for "such court orders and legal process shall be satisfied".

Subsec. (e)(5). Pub. L. 98-525, Sec. 643(d)(4), struck out "of disposable retired or retainer pay" after "payment of an amount" in two places and substituted "disposable retired or retainer pay" for "such pay" before "available for payment".

EFFECTIVE DATE OF 1993 AMENDMENT

Section 555(c) of Pub. L. 103-160 provided that: "The amendments made by this section (amending this section) shall take effect as of October 23, 1992, and shall apply as if the provisions of the paragraph (10) of section 1408(h) of title 10, United States Code, added by such subsection were included in the amendment made by section 653(a)(2) of Public Law 102-484 (106 Stat. 2426) (amending this section)."

EFFECTIVE DATE OF 1990 AMENDMENT

Section 555(e) of Pub. L. 101-510, as amended by Pub. L. 102-190, div. A, title X, Sec. 1062(a)(1), Dec. 5, 1991, 105 Stat. 1475, provided that:

"(1) The amendment made by subsection (a) (amending this section) shall apply with respect to judgments issued before, on, or after the date of the enactment of this Act (Nov. 5, 1990). In the case of a judgment issued before the date of the enactment of this Act, such amendment shall not relieve any obligation, otherwise valid, to make a payment that is due to be made before the end of the two-year period beginning on the date of the enactment of this Act.

"(2) The amendments made by subsections (b), (c), and (d) (amending this section) apply with only respect to divorces, dissolutions of marriage, annulments, and legal separations that become effective after the end of the 90-day period beginning on the date of the enactment of this Act."

EFFECTIVE DATE OF 1987 AMENDMENT

Amendment by section 3(3) of Pub. L. 100-26 applicable as if included in Pub. L. 99-661 when enacted on Nov. 14, 1986, see section 12(a) of Pub. L. 100-26, set out as a note under section 776 of this title.

EFFECTIVE DATE OF 1986 AMENDMENT

Section 644(b) of Pub. L. 99-661 provided that: "The amendments made by subsection (a) (amending this section) shall apply with respect to court orders issued after the date of the enactment of this Act (Nov. 14, 1986)."

EFFECTIVE DATE OF 1984 AMENDMENT

Section 643(e) of Pub. L. 98-525 provided that: "The amendments made by this section (amending this section) shall apply with respect to court orders for which effective service (as described in section 1408(b)(1) of title 10, United States Code, as amended by subsection (b) of this section) is made on or after the date of the enactment of this Act (Oct. 19, 1984)."

EFFECTIVE DATE; TRANSITION PROVISIONS

Section 1006 of title X of Pub. L. 97-252, as amended by Pub. L. 98-94, title IX, Sec. 941(c)(4), Sept. 24, 1983, 97 Stat. 654; Pub. L. 98-525, title VI, Sec. 645(b), Oct. 19, 1984, 98 Stat. 2549, provided that:

"(a) The amendments made by this title (amending this section and sections 1072, 1076, 1086, 1447, 1448, and 1450 of this title and enacting provisions set out as notes under this section and section 1408 of this title) shall take effect on the first day of the first month (February 1983) which begins more than one hundred and twenty days after the date of the enactment of this title (Sept. 8, 1982).

"(b) Subsection (d) of section 1408 of title 10, United States Code, as added by section 1002(a), shall apply only with respect to payments of retired or retainer pay for periods beginning on or after the effective date of this title (Feb. 1, 1983, provided in subsec. (a)), but without regard to the date of any court order. However, in the case of a court order that became final before June 26, 1981, payments under such subsection may only be made in accordance with such order as in effect on such date and without regard to any subsequent modifications.

"(c) The amendments made by section 1003 of this title (amending sections 1447, 1448, and 1450 of this title) shall apply to persons who become eligible to participate in the Survivor Benefit Plan provided for in subchapter II of chapter 73 of title 10, United States Code (section 1447 et seq. of this title), before, on, or after the effective date of such amendments.

"(d) The amendments made by section 1004 of this title (amending sections 1072, 1076, and 1086 of this title) and the provisions of section 1005 of this title (set out as a note under this section) shall apply in the case of any former spouse of a member or former member of the uniformed services whether the final decree of divorce, dissolution, or annulment of the marriage of the former spouse and such member or former member is dated before, on, or after February 1, 1983.

"(e) For the purposes of this section –

"(1) the term 'court order' has the same meaning as provided in section 1408(a)(2) of title 10, United States Code (as added by section 1002 of this title);

"(2) the term 'former spouse' has the same meaning as

provided in section 1408(a)(6) of such title (as added by section 1002 of this title); and

"(3) the term 'uniformed services' has the same meaning as provided in section 1072 of title 10, United States Code."

TERMINATION OF TRUST TERRITORY OF THE PACIFIC ISLANDS

For termination of Trust Territory of the Pacific Islands, see note set out preceding section 1681 of Title 48, Territories and Insular Possessions.

ACCRUAL OF PAYMENTS; PROSPECTIVE APPLICABILITY

Section 653(c) of Pub. L. 102-484 provided that: "No payments under subsection (h) of section 1408 of title 10, United States Code (as added by subsection (a)), shall accrue for periods before the date of the enactment of this Act (Oct. 23, 1992)."

STUDY CONCERNING BENEFITS FOR DEPENDENTS WHO ARE VICTIMS OF ABUSE

Section 653(e) of Pub. L. 102-484 provided that:

"(1) The Secretary of Defense shall conduct a study in order to estimate –

"(A) the number of persons who will become eligible to receive payments under subsection (h) of section 1408 of title 10, United States Code (as added by subsection (a)), during each of fiscal years 1993 through 2000; and

"(B) for each of fiscal years 1993 through 2000, the number of members of the Armed Forces who, after having completed at least one, and less than 20, years of service in that fiscal year, will be approved in that fiscal year for separation from the Armed Forces as a result of having abused a spouse or dependent child.

"(2) The study shall include a thorough analysis of –

"(A) the effects, if any, of appeals and requests for clemency

144

in the case of court-martial convictions on the entitlement to payments in accordance with subsection (h) of section 1408 of title 10, United States Code (as added by subsection (a));

"(B) the socio-economic effects on the dependents of members of the Armed Forces described in subsection (h)(2) of such section that result from terminations of the eligibility of such members to receive retired or retainer pay; and

"(C) the effects of separations of such members from the Armed Forces on the mission readiness of the units of assignment of such members when separated and on the Armed Forces in general.

"(3) Not later than one year after the date of the enactment of this Act (Oct. 23, 1992), the Secretary shall submit to Congress a report on the results of the study."

COMMISSARY AND EXCHANGE PRIVILEGES

Section 1005 of Pub. L. 97-252, which directed Secretary of Defense to prescribe regulations to provide that an unremarried former spouse described in 10 U.S.C. 1072(2)(F)(i) is entitled to commissary and post exchange privileges to the same extent and on the same basis as the surviving spouse of a retired member of the uniformed services, was repealed and restated in section 1062 of this title by Pub. L. 100-370, Sec. 1(c)(1), (5).

SECTION REFERRED TO IN OTHER SECTIONS

This section is referred to in sections 1059, 1078a, 1447, 1461, 1463 of this title.

THE DECISION

McCARTY v. McCARTY, 453 U.S. 210 (1981)

453 U.S. 210

McCARTY v. McCARTY.
APPEAL FROM THE COURT OF APPEAL OF
CALIFORNIA, FIRST APPELLATE DISTRICT.

No. 80-5.

Argued March 2, 1981.
Decided June 26, 1981.

A regular commissioned officer of the United States Army who retires after 20 years of service is entitled to retired pay. Retired pay terminates with the officer's death, although he may designate a beneficiary to receive any arrearages that remain unpaid at death. In addition there are statutory plans that allow the officer to set aside a portion of his retired pay for his survivors. Appellant, a Regular Army Colonel, filed a petition in California Superior Court for dissolution of his marriage to appellee. At the time, he had served approximately 18 of the 20 years required for retirement with pay. Under California law, each spouse, upon dissolution of a marriage, has an equal and absolute right to a half interest in all community and quasi-community property, but retains his or her separate property. In his petition, appellant requested, inter alia, that his military retirement benefits be confirmed to him as his separate property. The Superior Court held, however, that such benefits were subject to division as quasi-community property, and accordingly ordered appellant to pay to appellee a specified portion of the benefits upon retirement. Subsequently, appellant retired and began receiving retired pay; under the dissolution decree, appellee was entitled to approximately 45% of the retired pay. On review of this award, the California Court of Appeal affirmed, rejecting appellant's contention that because the federal scheme of military retirement benefits pre-empts state community property law, the Supremacy Clause precluded the trial court from awarding appellee a portion of his retired pay.

Held:

Federal law precludes a state court from dividing military retired pay pursuant to state community property laws. Pp. 220-236.

(a) There is a conflict between the terms of the federal military retirement statutes and the community property right asserted by appellee. The military retirement system confers no entitlement to retired pay upon the retired member's spouse, and does not embody even a limited "community property concept." Rather, the language, structure, and history of the statutes make it clear that retired pay continues to be the personal entitlement of the retiree. Pp. 221-232.

(b) Moreover, the application of community property principles to military retired pay threatens grave harm to "clear and substantial" [453 U.S. 210, 211] federal interests. Thus, the community property division of retired pay, by reducing the amounts that Congress has determined are necessary for the retired member, has the potential to

147

frustrate the congressional objective of providing for the retired service member. In addition, such a division has the potential to interfere with the congressional goals of having the military retirement system serve as an inducement for enlistment and re-enlistment and as an encouragement to orderly promotion and a youthful military. Pp. 232-235.

Reversed and remanded.

BLACKMUN, J., delivered the opinion of the Court, in which BURGER, C. J., and WHITE, MARSHALL, POWELL, and STEVENS, JJ., joined. REHNQUIST, J., filed a dissenting opinion, in which BRENNAN and STEWART, JJ., joined, post, p. 236.

Mattaniah Eytan argued the cause and filed briefs for appellant.

Walter T. Winter argued the cause for appellee. With him on the brief was Barbara R. Dornan. *

[Footnote *] Herbert N. Harmon filed a brief for the Non-Commissioned Officers Association of the United States of America et al. as amici curiae urging reversal.

Briefs of amici curiae urging affirmance were filed by William H. Allen for John L. Burton et al.; and by Gertrude D. Chern, Judith I. Avner, Gill Deford, and Neal Dudovitz for the National Organization for Women Legal Defense and Education Fund et al.

JUSTICE BLACKMUN delivered the opinion of the Court.

A regular or reserve commissioned officer of the United States Army who retires after 20 years of service is entitled to retired pay. 10 U.S.C. 3911 and 3929. The question presented by this case is whether, upon the dissolution of a marriage, federal law precludes a state court from dividing military non-disability retired pay pursuant to state community property laws.

I

Although disability pensions have been provided to military veterans from the Revolutionary War period to the [453 U.S. 210, 212] present, 1 it was not until the War Between the States that Congress enacted the first comprehensive non-disability military retirement legislation. See Preliminary Review of Military Retirement Systems: Hearings before the Military Compensation Subcommittee of the House Committee on Armed Services, 95th Cong., 1st and Sess., 5 (1977-1978) (Military Retirement Hearings) (statement of Col. Leon S. Hirsh, Jr., USAF, Director of Compensation, Office of the Assistant Secretary of Defense for Manpower, Reserve Affairs, and Logistics); Subcommittee on Retirement Income and Employment, House Select Committee on Aging, Women and Retirement Income Programs: Current Issues of Equity and Adequacy, 96th Cong., 1st Sess., 15 (Comm. Print 1979) (Women and Retirement). Sections 15 and 21 of the Act of Aug. 3, 1861, 12 Stat. 289, 290, provided that any Army, Navy, or Marine Corps officer with 40 years of service could apply to the President to be retired with pay; in addition, 16 and 22 of that Act authorized the involuntary retirement with pay of any officer "incapable of performing the duties of his office." 12 Stat. 289, 290.

The impetus for this legislation was the need to encourage or force the retirement of officers who were not fit for wartime duty. 2 Women and Retirement, at 15. Thus, from [453 U.S. 210, 213] its inception, 3 the military non-disability retirement system has been "as much a personnel management tool as an income maintenance method," id., at 16; the system was and is designed not only to provide for retired officers, but also to ensure a "young and vigorous" military force, to create an orderly pattern of promotion, and to serve as a recruiting and re-enlistment inducement. Military Retirement Hearings, at 4-6, 13 (statement of Col. Hirsh).

Under current law, there are three basic forms of military retirement: non-disability retirement; disability retirement; and reserve retirement. See id., at 4. For our present purposes, only the first of these three forms is relevant. 4 Since each of the military services has substantially the same non-disability retirement system, see id., at 5, the Army's system may be taken as typical. 5 An Army officer who has 20 years of service, at least 10 of which have been active service as a commissioned officer, may request that the Secretary of the [453 U.S. 210, 214] Army retire him. 10 U.S.C. 3911. 6 An officer who requests such retirement is entitled to "retired pay." This is calculated on the basis of the number of years served and rank achieved. 3929 and 3991. 7 An officer who serves for less than 20 years is not entitled to retired pay.

The non-disability retirement system is noncontributory in that neither the service member nor the Federal Government makes periodic contributions to any fund during the period of active service; instead, retired pay is funded by annual

150

appropriations. Military Retirement Hearings, at 5. In contrast, since 1957, military personnel have been required to contribute to the Social Security System. Pub. L. 84-881, 70 Stat. 870. See 42 U.S.C. 410 (l) and (m). Upon satisfying the necessary age requirements, the Army retiree, the [453 U.S. 210, 215] spouse. an ex-spouse who was married to the retiree for at least 10 years, and any dependent children are entitled to Social Security benefits. See 42 U.S.C. 402 (a) to (f) (1976 ed. and Supp. IV).

Military retired pay terminates with the retired service member's death, and does not pass to the member's heirs. The member, however, may designate a beneficiary to receive any arrearages that remain unpaid at death. 10 U.S.C. 2771. In addition, there are statutory schemes that allow a service member to set aside a portion of the member's retired pay for his or her survivors. The first such scheme, now known as the Retired Serviceman's Family Protection Plan (RSFPP), was established in 1953. Act of Aug. 8, 1953, 67 Stat. 501, current version at 10 U.S.C. 1431-1446 (1976 ed. and Supp. IV). Under the RSFPP, the military member could elect to reduce his or her retired pay in order to provide, at death, an annuity for a surviving spouse or child. Participation in the RSFPP was voluntary, and the participating member, prior to receiving retired pay, could revoke the election in order "to reflect a change in the marital or dependency status of the member or his family that is caused by death, divorce, annulment, remarriage, or acquisition of a child . . . ." 1431 (c). Further, deductions from retired pay automatically cease upon the death or divorce of the service member's spouse. 1434 (c).

Because the RSFPP was self-financing, it required the

151

deduction of a substantial portion of the service member's retired pay; consequently, only about 15% of eligible military retirees participated in the plan. See H. R. Rep. No. 92-481. pp. 4-5 (1971); S. Rep. No. 92-1089. p. 11 (1972). In order to remedy this situation. Congress enacted the Survivor Benefit Plan (SBP) in 1972. Pub. L. 92-425. 86 Stat. 706, codified, as amended, at 10 U.S.C. 1447-1455 (1976 ed. and Supp. IV). Participation in this plan is automatic unless the service member chooses to opt out. 1448 (a). [453 U.S. 210, 216] The SBP is not entirely self-financing; instead, the Government contributes to the plan, thereby rendering participation in the SBP less expensive for the service member than participation in the RSFPP. Participants in the RSFPP were given the option of continuing under that plan or of enrolling in the SBP. Pub. L. 92-425, 3, 86 Stat. 711, as amended by Pub. L. 93-155, 804, 87 Stat. 615.

II

Appellant Richard John McCarty and appellee Patricia Ann McCarty were married in Portland, Ore., on March 23, 1957, while appellant was in his second year in medical school at the University of Oregon. During his fourth year in medical school, appellant commenced active duty in the United States Army. Upon graduation, he was assigned to successive tours of duty in Pennsylvania, Hawaii, Washington, D.C., California, and Texas. After completing his duty in Texas, appellant was assigned to Letterman Hospital on the Presidio Military Reservation in San Francisco, where he became Chief of Cardiology. At the time this suit was instituted in 1976, appellant held the rank of Colonel and had served approximately 18 of the 20 years required under 10 U.S.C. 3911 for retirement with pay.

Appellant and appellee separated on October 31, 1976. On December 1 of that year, appellant filed a petition in the Superior Court of California in and for the City and Country of San Francisco requesting dissolution of the marriage. Under California law, a court granting dissolution of a marriage must divide "the community property and the quasi-community property of the parties." Cal. Civ. Code Ann. 4800 (a) (West Supp. 1981). Like seven other States, California treats all property earned by either spouse during the marriage as community property; each spouse is deemed to make an equal contribution to the marital enterprise, and therefore each is entitled to share equally in its assets. See [453 U.S. 210, 217]   Hisquierdo v. Hisquierdo, <u>439 U.S. 572, 577</u> -578 (1979). "Quasi-community property" is defined as

"all real or personal property, wherever situated heretofore or hereafter acquired . . . [b]y either spouse while domiciled elsewhere which would have been community property if the spouse who acquired the property had been domiciled in [California] at the time of its acquisition." Cal. Civ. Code Ann. 4803 (West Supp. 1981).

Upon dissolution of a marriage, each spouse has an equal and absolute right to a half interest in all community and quasi-community property; in contrast, each spouse retains his or her separate property, which includes assets the spouse owned before marriage or acquired separately during marriage through gift. See Hisquierdo, <u>439 U.S., at 578</u> .

In his dissolution petition, appellant requested that all listed assets, including "[a]ll military retirement benefits," be confirmed to him as his separate property. App. 2. In her response, appellee also requested dissolution of the marriage, but contended that appellant had no separate property and that therefore his military retirement benefits were "subject to disposition by the court in this proceeding." 8 Id., at 8-9. On November 23, 1977, the Superior Court entered findings of fact and conclusions of law holding that appellant was entitled to an interlocutory judgment dissolving [453 U.S. 210, 218] the marriage. Id., at 39, 44. Appellant was awarded custody of the couple's three minor children; appellee was awarded spousal support. The court found that the community property of the parties consisted of two automobiles, cash, the cash value of life insurance policies, and an uncollected debt. Id., at 42. It allocated this property between the parties. Id., at 45. In addition, the court held that appellant's "military pension and retirement rights" were subject to division as quasi-community property. Ibid. Accordingly, the court ordered appellant to pay to appellee, so long as she lives,

"that portion of his total monthly pension or retirement payment which equals one-half (1/2) of the ratio of the total time between marriage and separation during which [appellant] was in the United States Army to the total number of years he has served with the . . . Army at the time of retirement." Id., at 43-44.

The court retained jurisdiction "to make such determination at that time and to supervise distribution . . . ." Ibid. On September 30, 1978, appellant retired from the Army after 20 years of active duty and began receiving retired pay; under the decree of dissolution, appellee was entitled to approximately 45% of that retired pay.

154

Appellant sought review of the portion of the Superior Court's decree that awarded appellee an interest in the retired pay. The California Court of Appeal, First Appellate District, however, affirmed the award. App. to Juris. Statement 32. In so ruling, the court declined to accept appellant's contention that because the federal scheme of military retirement benefits pre-empts state community property laws, the Supremacy Clause, U.S. Const., Art. VI, cl. 2, precluded the trial court from awarding appellee a portion of his retired pay. 9 The court noted that this precise contention had [453 U.S. 210, 219] been rejected in In re Fithian, 10 Cal. 3d 592, 517 P.2d 449, cert. denied, 419 U.S. 825 (1974). 10 Furthermore, the court concluded that the result in Fithian had not been called into question by this Court's subsequent decision in Hisquierdo v. Hisquierdo, supra, where it was held that benefits payable under the federal Railroad Retirement Act of 1974 could not be divided under state community property law. See also Gorman v. Gorman, 90 Cal. App. 3d 454, 153 Cal. Rptr. 479 (1979). 11

The California Supreme Court denied appellant's petition for hearing. App. to Juris. Statement 83.

We postponed jurisdiction. 449 U.S. 917 (1980). We have now concluded that this case properly falls within our appellate jurisdiction, 12 and we therefore proceed to the merits. [453 U.S. 210, 220]

III

This Court repeatedly has recognized that "`[t]he whole subject of the domestic relations of husband and wife . . . belongs to the laws of the States and not to the laws of the

155

United States.'" Hisquierdo, <u>439 U.S., at 581</u>, quoting In re Burrus, <u>136 U.S. 586, 593</u> -594 (1890). Thus, "[s]tate family and family-property law must do `major damage' to `clear and substantial' federal interests before the Supremacy Clause will demand that state law be overridden." Hisquierdo, <u>439 U.S., at 581</u>, with references to United States v. Yazell, <u>382 U.S. 341, 352</u> (1966). See also Alessi v. Raybestos-Manhattan, Inc., <u>451 U.S. 504, 522</u> (1981). In Hisquierdo, we concluded that California's application of community property principles to Railroad Retirement Act benefits worked such an injury to federal interests. The "critical terms" of the federal statute relied upon in reaching that conclusion included provisions establishing "a specified beneficiary protected by a flat prohibition against attachment and anticipation," see 45 U.S.C. 231m, and a limited community property concept that terminated upon divorce, see 45 U.S.C. 231d. <u>439 U.S., at 582</u> -585. Appellee argues that no such provisions are to be found in the statute presently under consideration, and that therefore Hisquierdo is inapposite. But Hisquierdo did not hold that only the particular statutory terms there considered would justify a finding [453 U.S. 210, 221] of pre-emption; rather, it held that "[t]he pertinent questions are whether the right as asserted conflicts with the express terms of federal law and whether its consequences sufficiently injure the objectives of the federal program to require nonrecognition." Id., at 583. It is to that twofold inquiry that we now turn.

A

Appellant argues that California's application of community property concepts to military retired pay conflicts with federal law in two distinct ways. He contends, first, that the

California court's conclusion that retired pay is "awarded in return for services previously rendered," see Fithian, 10 Cal. 3d, at 604, 517 P.2d, at 457, ignores clear federal law to the contrary. The community property division of military retired pay rests on the premise that that pay, like a typical pension, represents deferred compensation for services performed during the marriage. Id., at 596, 517 P.2d, at 451. But, appellant asserts, military retired pay in fact is current compensation for reduced, but currently rendered, services; accordingly, even under California law, that pay may not be treated as community property to the extent that it is earned after the dissolution of the marital community, since the earnings of a spouse while living "separate and apart" are separate property. Cal. Civ. Code Ann. 5118, 5119 (West 1970 and Supp. 1981).

Appellant correctly notes that military retired pay differs in some significant respects from a typical pension or retirement plan. The retired officer remains a member of the Army, see United States v. Tyler, <u>105 U.S. 244</u> (1882), 13 and [453 U.S. 210, 222] continues to be subject to the Uniform Code of Military Justice, see 10 U.S.C. 802 (4). See also Hooper v. United States, 164 Ct. Cl. 151, 326 F.2d 982, cert. denied, <u>377 U.S. 977</u> (1964). In addition, he may forfeit all or part of his retired pay if he engages in certain activities. 14 Finally, the retired officer remains subject to recall to active duty by the Secretary of the Army "at any time." Pub. L. 96-513, 106, 94 Stat. 2868. These factors have led several courts, including this one, to conclude that military retired pay is reduced compensation for reduced current services. In United States v. Tyler, <u>105 U.S., at 245</u>, the Court stated that retired pay is "compensation . . . continued at a reduced

rate, and the connection is continued, with a retirement from active service only." 15 [453 U.S. 210, 223]

Having said all this, we need not decide today whether federal law prohibits a State from characterizing retired pay as deferred compensation, since we agree with appellant's alternative argument that the application of community property law conflicts with the federal military retirement scheme regardless of whether retired pay is defined as current or as deferred compensation. 16 The statutory language is straight-forward: [453 U.S. 210, 224] "A member of the Army retired under this chapter is entitled to retired pay . . . ." 10 U.S.C. 3929. In Hisquierdo, <u>439 U.S., at 584</u> , we emphasized that under the Railroad Retirement Act a spouse of a retired railroad worker was entitled to a separate annuity that terminated upon divorce, see 45 U.S.C. 231d (c) (3); in contrast, the military retirement system confers no entitlement to retired pay upon the retired service member's spouse. Thus, unlike the Railroad Retirement Act, the military retirement system does not embody even a limited "community property concept." Indeed, Congress has explicitly stated: "Historically, military retired pay has been a personal entitlement payable to the retired member himself as long as he lives." S. Rep. No. 1480, 90th Cong., 2d Sess., 6 (1968) (emphasis added).

Appellee argues that Congress' use of the term "personal entitlement" in this context signifies only that retired pay ceases upon the death of the service member. But several features of the statutory schemes governing military pay demonstrate that Congress did not use the term in so limited a fashion. First, the service member may designate a beneficiary to receive any unpaid arrearages in retired pay upon his death. 10 U.S.C. 2771. 17 The service member is free [453 U.S. 210, 225] to designate someone other than his spouse or ex-spouse as the beneficiary; further, the statute expressly provides that "[a] payment under this section bars recovery by any other person of the amount paid." 2771 (d). In Wissner v. Wissner, 338 U.S. 655 (1950), this Court considered an analogous statutory scheme. Under the National Service Life Insurance Act, an insured service member had the right to designate the beneficiary of his policy. Id., at 658. Wissner held that California could not award a service member's widow half the proceeds of a life insurance policy, even though the source of the premiums - the member's Army pay - was characterized as community property under California law. The Court reserved the question whether California is "entitled to call army pay community property," id., at 657, n. 2, since it found that Congress had "spoken with force and clarity in directing that the proceeds belong to the named beneficiary and no other." Id., at 658. In the present context, Congress has stated with "force and clarity" that a beneficiary under 2771 claims an interest in the retired [453 U.S. 210, 226] pay itself, not simply in proceeds from a policy purchased with that pay. One commentator has noted: "If retired pay were community property, the retiree could not thus summarily deprive his wife of her interest in the arrearage." Goldberg, Is Armed Services Retired Pay Really Community Property?, 48 Cal. Bar J. 12, 17 (1973).

Second, the language, structure, and legislative history of the RSFPP and the SBP also demonstrate that retired pay is a "personal entitlement." While retired pay ceases upon the death of the service member, the RSFPP and the SBP allow the service member to reduce his or her retired pay in order to provide an annuity for the surviving spouse or children. Under both plans, however, the service member is free to elect to provide no annuity at all, or to provide an annuity payable only to the surviving children, and not to the spouse. See 10 U.S.C. 1434 (1976 ed. and Supp. IV) (RSFPP); 1450 (1976 ed. and Supp. IV) (SBP). Here again, it is clear that if retired pay were community property, the service member could not so deprive the spouse of his or her interest in the property. 18 But we need not rely on this implicit conflict alone, for both the language of the statutes 19 and their legislative history make it clear that the [453 U.S. 210, 227] decision whether to leave an annuity is the service member's decision alone because retired pay is his or her personal entitlement. It has been stated in Congress that "[t]he rights in retirement pay accrue to the retiree and, ultimately, the decision is his as to whether or not to leave part of that retirement pay as an annuity to his survivors." H. R. Rep. No. 92-481, p. 9 (1971). 20 California's community property division of retired pay is simply inconsistent with this explicit expression of congressional intent that retired pay accrue to the retiree.

Moreover, such a division would have the anomalous effect of placing an ex-spouse in a better position than that of a widower or a widow under the RSFPP and the SBP. 21 Appellee [453 U.S. 210, 228]   argues that "Congress' concern for the welfare of soldiers' widows sheds little light on Congress' attitude toward the community treatment of retirement benefits," quoting Fithian, 10 Cal. 3d, at 600, 517 P.2d, at 454. But this argument fails to recognize that Congress deliberately has chosen to favor the widower or widow over the ex-spouse. An ex-spouse is not an eligible beneficiary of an annuity under either plan. 10 U.S.C. 1434

(a) (RSFPP); 1447 (3) and 1450 (a) (SBP). In addition, under the RSFPP, deductions from retired pay for a spouse's annuity automatically cease upon divorce, 1434 (c), so as "[t]o safeguard the participants' future retired pay when . . . divorce occurs . . . ." S. Rep. No. 1480, 90th Cong., 2d Sess., 13 (1968). While the SBP does not expressly provide that annuity deductions cease upon divorce, the legislative history indicates that Congress' policy remained unchanged. The SBP, which was referred to as the "widow's equity bill," 118 Cong. Rec. 29811 (1972) (statement of Sen. Beall), was enacted because of Congress' concern over the number of widows left without support through low participation in the RSFPP, not out of concern for ex-spouses. See H. R. Rep. No. 92-481, pp. 4-5 (1971); S. Rep. No. 92-1089, p. 11 (1972).

Third, and finally, it is clear that Congress intended that military retired pay "actually reach the beneficiary." See Hisquierdo, 439 U.S., at 584 . Retired pay cannot be attached to satisfy a property settlement incident to the dissolution of a marriage. 22 In enacting the SBP, Congress rejected [453 U.S. 210, 229] a provision in the House bill, H. R. 10670, that would have allowed attachment of up to 50% of military retired pay to comply with a court order in favor of a spouse, former spouse, or child. See H. R. Rep. No. 92-481, at 1; S. Rep. No. 92-1089, at 25. The House Report accompanying H. R. 10670 noted that under Buchanan v. Alexander, 4 How. 20 (1845), and Applegate v. Applegate, 39 F. Supp. 887 (ED Va. 1941), military pay could not be attached so long as it was in the Government's hands; 23 thus, this clause of H. R. 10670 represented a "drastic departure" from current law, but one that the House Committee on Armed Services believed to be necessitated by the difficulty of enforcing support orders. H. R. Rep. No. 92-481, at 17-18. Although this provision passed the House, it was not included in the Senate version of the bill. See S. Rep. No. 92-1089, at 25. Thereafter, the House acceded to the Senate's view that the attachment provision would unfairly "single out military retirees for a form of enforcement of court orders imposed on no other employees or retired employees of the Federal Government." 118 Cong. Rec. 30151 (1972) (remarks of Rep. Pike); S. Rep. No. 92-1089, [453 U.S. 210, 230] at 25. Instead, Congress determined that the problem of the attachment of military retired pay should be considered in the context of "legislation that might require all Federal pays to be subject to attachment." Ibid.; 118 Cong. Rec. 30151 (1972) (remarks of Rep. Pike).

Subsequently, comprehensive legislation was enacted. In 1975, Congress amended the Social Security Act to provide that all federal benefits, including those payable to members of the Armed Services, may be subject to legal process to enforce child support or alimony obligations. Pub. L. 93-647, 101 (a), 88 Stat. 2357, 42 U.S.C. 659. In 1977, however, Congress added a new definitional section (462 (c)) providing that the term "alimony" in 659 (a) "does not include any payment or transfer of property . . . in compliance with any community property settlement, equitable distribution of property, or other division of property between spouses or former spouses." Pub. L. 95-30, 501 (d), 91 Stat. 159, 42 U.S.C. 662 (c) (1976 ed., Supp. IV). As we noted in Hisquierdo, it is "logical to conclude that Congress, in adopting 462 (c), thought that a family's need for support could justify garnishment, even though it deflected other federal benefits from their intended goals, but that community property claims, which are not based on need, could not do so." 439 U.S., at 587 .

Hisquierdo also pointed out that Congress might conclude that this distinction between support and community property claims is "undesirable." Id., at 590. Indeed, Congress recently enacted legislation that requires that Civil Service retirement benefits be paid to an ex-spouse to the extent provided for in "the terms of any court order or court-approved property settlement agreement incident to any court decree of divorce, annulment, or legal separation." Pub. L. 95-366, 1 (a), 92 Stat. 600, 5 U.S.C. 8345 (j) (1) (1976 ed., Supp. IV). In an even more extreme recent step, Congress amended the Foreign Service retirement legislation to provide that, as a matter of federal law, an ex-spouse is entitled [453 U.S. 210, 231]   to a pro rata share of Foreign Service retirement benefits. 24 Thus, the Civil Service amendments require the United States to recognize the community property division of Civil Service retirement benefits by a state court, while the Foreign Service amendments establish a limited federal community property concept. Significantly, however, while similar legislation affecting military retired pay was introduced in the 96th Congress, none of those bills was reported out of committee. 25 Thus, in striking contrast to its amendment [453 U.S. 210, 232]   of the Foreign Service and Civil Service retirement systems, Congress has neither authorized nor required the community property division of military retired pay. On the contrary, that pay continues to be the personal entitlement of the retiree.

B

We conclude, therefore, that there is a conflict between the terms of the federal retirement statutes and the community property right asserted by appellee here. But "[a] mere conflict in words is not sufficient"; the question remains whether the "consequences [of that community property right] sufficiently injure the objectives of the federal program to require nonrecognition." Hisquierdo, 439 U.S., at 581 -583. This inquiry, however, need be only a brief one, for it is manifest that the application of community property principles to military retired pay threatens grave harm to "clear and substantial" federal interests. See United States v. Yazell, 382 U.S., at 352 . Under the Constitution, Congress has the power "[t]o raise and support Armies," "[t]o provide and maintain a Navy," and "[t]o makes Rules for the Government and Regulation of the land and naval Forces." U.S. Const., Art. I, 8, cls. 12, 13, and 14. See generally Rostker v. Goldberg, ante, at 59. Pursuant to this grant of authority, Congress has enacted a military retirement system designed to accomplish two major goals: to provide for the retired service member, and to meet the personnel management [453 U.S. 210, 233] needs of the active military forces. The community property division of retired pay has the potential to frustrate each of these objectives.

In the first place, the community property interest appellee seeks "promises to diminish that portion of the benefit Congress has said should go to the retired [service member] alone." See Hisquierdo, 439 U.S., at 590 . State courts are not free to reduce the amounts that Congress has determined are necessary for the retired member. Furthermore, the community property division of retired pay may disrupt the carefully balanced scheme Congress has devised to encourage a service member to set aside a portion of his or her retired pay as an annuity for a surviving spouse or dependent children. By diminishing the amount available to the retiree, a community property division makes it less likely that the retired service member will choose to reduce his or her retired pay still further by purchasing an annuity for the surviving spouse, if any, or children. In McCune v. Essig, 199 U.S. 382 (1905), the Court held that federal law, which permitted a widow to patent federal land entered by her husband, prevailed over the interest in the patent asserted by the daughter under state inheritance law; the Court noted that the daughter's contention "reverses the order of the statute and gives the children an interest Paramount to that of the widow through the laws of the State." Id., at 389. So here, the right appellee asserts "reverses the order of the statute" by giving the ex-spouse an interest paramount to that of the surviving spouse and children of the service member; indeed, at least one court (in a noncommunity property State) has gone so far as to hold that the heirs of the ex-spouse may even inherit her interest in military retired pay. See In re Miller, ___ Mont. ___, 609 P.2d 1185 (1980), cert. pending sub nom. Miller v. Miller, No. 80-291. Clearly, "[t]he law of the State is not competent to do this." McCune v. Essig, 199 U.S., at 389. [453 U.S. 210, 234]

The potential for disruption of military personnel management is equally clear. As has been noted above, the military retirement system is designed to serve as an inducement for enlistment and re-enlistment, to create an orderly career path, and to ensure "youthful and vigorous" military forces. 26 While conceding that there is a substantial interest in attracting and retaining personnel for the military forces, appellee argues that this interest will not be impaired by allowing a State to apply its community property laws to retired military personnel in the same manner that it applies those laws to civilians. Yet this argument ignores two essential characteristics of military service: the military forces are national in operation; and their members, unlike civilian employees, cf. Hisquierdo, are not free to choose their place of residence. Appellant, for instance, served tours of duty in four States and the District of Columbia. The value of retired pay as an inducement for enlistment or re-enlistment is obviously diminished to the extent that the service member recognizes that he or she may be involuntarily transferred to a State that will divide that pay upon divorce. In Free v. Bland, [453 U.S. 210, 235] 369 U.S. 663 (1962), the Court held that state community property law could not override the survivorship provision of a federal savings bond, since it was "[o]ne of the inducements selected," id., at 669, to make purchase of such bonds attractive; similarly, retired pay is one of the inducements selected to make military service attractive, and the application of state community property law thus "interfere[s] directly with a legitimate exercise of the power of the Federal Government." Ibid.

The interference with the goals of encouraging orderly promotion and a youthful military is no less direct. Here, as in the Railroad Retirement Act context, "Congress has fixed an amount thought appropriate to support an employee's old age and to encourage the employee to retire." See Hisquierdo, 439 U.S., at 585 . But the reduction of retired pay by a community property award not only discourages retirement by reducing the retired pay available to the service member, but gives him a positive incentive to keep working, since current income after divorce is not divisible as community property. See Cal. Civ. Code Ann. 5118, 5119 (West 1970 and Supp. 1981). Congress has determined that a

166

youthful military is essential to the national defense; it is not for States to interfere with that goal by lessening the incentive to retire created by the military retirement system.

IV

We recognize that the plight of an ex-spouse of a retired service member is often a serious one. See Hearing on H. R. 2817, H. R. 3677, and H. R. 6270 before the Military Compensation Subcommittee of the House Committee on Armed Services, 96th Cong., 2d Sess. (1980). That plight may be mitigated to some extent by the ex-spouse's right to claim Social Security benefits, cf. Hisquierdo, 439 U.S., at 590 , and to garnish military retired pay for the purposes of support. Nonetheless, Congress may well decide, as it has in the Civil Service and Foreign Service contexts, that more protection [453 U.S. 210, 236]   should be afforded a former spouse of a retired service member. This decision, however, is for Congress alone. We very recently have re-emphasized that in no area has the Court accorded Congress greater deference than in the conduct and control of military affairs. See Rostker v. Goldberg, ante, at 64-65. Thus, the conclusion that we reached in Hisquierdo follows a fortiori here: Congress has weighed the matter, and "[i]t is not the province of state courts to strike a balance different from the one Congress has struck." 439 U.S., at 590 .

The judgment of the California Court of Appeal is reversed, and the case is remanded for further proceedings not inconsistent with this opinion. \ It is so ordered.

Footnotes

[Footnote 1] See Rombauer, Marital Status and Eligibility for Federal Statutory Income Benefits: A Historical Survey, 52 Wash. L. Rev. 227, 228-229 (1977). The current military disability provisions are 10 U.S.C. 1201 et seq. (1976 ed. and Supp. IV).

[Footnote 2] See Cong. Globe, 37th Cong., 1st Sess., 16 (1861) (remarks of Sen. Grimes) ("some of the commanders of regiments in the regular service are utterly incapacitated for the performance of their duty, and they ought to be retired upon some terms, and efficient men placed in their stead"); id., at 159 (remarks of Sen. Wilson) ("We have colonels, lieutenant colonels, and majors in the Army, old men, worn out by exposure in the service, who cannot perform their duties; men who ought to be honorably retired, and receive the compensation provided for in this measure").

[Footnote 3] For a survey of subsequent military non-disability legislation, see U.S. Dept. of Defense, Military Compensation Background Papers, Third Quadrennial Review of Military Compensation 183-202 (1976); Military Retirement Hearings, at 12-13.

[Footnote 4] For an overview of the disability and reserve retirement systems, see Subcommittee on Investigations, House Committee on Post Office and Civil Service, Dual Compensation Paid to Retired Uniformed Services' Personnel in Federal Civilian Positions, 95th Cong., 2d Sess., 18-20 (Comm. Print 1978).

[Footnote 5] The voluntary non-disability retirement

systems of the various services are codified as follows: 10 U.S.C., ch. 367, 3911 et seq. (1976 ed. and Supp. IV) (Army); ch. 571, 6321 et seq. (1976 ed. and Supp. IV) (Navy and Marine Corps); ch. 867, 8911 et seq. (Air Force). The non-disability retirement system was recently amended by the Defense Officer Personnel Management Act, Pub. L. 96-513, 94 Stat. 2835. Under 111 of that Act, id., at 2875, 10 U.S.C. 1251 (1976 ed., Supp. IV), regular commissioned officers in all the military services are required, with some exceptions, to retire at age 62; the Act also amended various provisions dealing with involuntary non-disability retirement for length of service. The Act, however, did not affect the particular voluntary non-disability retirement provisions at issue here.

[Footnote 6] An enlisted member of the Army may be retired upon his request after 30 years of service. 10 U.S.C. 3917. See also 3914, as amended by the Military Personnel and Compensation Amendments of 1980, Pub. L. 96-343, 9 (a) (1), 94 Stat. 1128, 10 U.S.C. 3914 (1976 ed., Supp. IV) (voluntary retirement after 20 years followed by service in Army Reserve). A retired enlisted member is also entitled to retired pay. 10 U.S.C. 3929 and 3991.

[Footnote 7] The amount of retired pay is calculated according to formula: (basic pay of the retired grade of the member) X (2 1/2%) X (the number of years of creditable service). Thus, a retiree is eligible for at least 50% (2 1/2% X 20 years of service) of his or her basic pay, which does not include special pay and allowances. There is, however, an upper limit of 75% of basic pay - the percentage attained upon retirement after completion of 30 years of service (30 years X 2 1/2%) - regardless of the number of years actually served. See 10 U.S.C. 3991. See generally Women and Retirement, at 16. The amount of retired pay is adjusted for any increase in the Consumer Price Index. 1401a.

Since the initiation of this suit, 3991 has been amended twice. See the Department of Defense Authorization Act, 1981, Pub. L. 96-342, 813 (c), 94 Stat. 1104, and the Defense Officer Personnel Management Act, Pub. L. 96-513, 502 (21), 94 Stat. 2910. Neither amendment has any bearing here.

Under the Internal Revenue Code of 1954, retired pay is taxable as ordinary income when received. 26 U.S.C. 61 (a) (11); 26 CFR 1.61-11 (1980).

[Footnote 8] At the time the interlocutory judgment of dissolution was entered, appellant had not begun to receive retired pay, since he had not yet completed 20 years of active service. Under California law, however, "pension rights" may be divided as community property even if they have not "vested." See In re Brown, 15 Cal. 3d 838, 544 P.2d 561 (1976). A California trial court may divide the present value of such rights, which value must take into account the possibility that death or termination of employment may destroy them before they vest. Id., at 848, 544 P.2d, at 567. Alternatively, the court may maintain continuing jurisdiction, and award each spouse an appropriate portion of each pension payment as it is made. Ibid. The trial court here apparently elected the latter alternative.

[Footnote 9] The Court of Appeal also held that since appellant had invoked the jurisdiction of the California courts over both his marital and property [453 U.S. 210, 219] rights, he was estopped from arguing that California community property law did not apply to him because he was an Oregon domiciliary. App. to Juris. Statement 50-54. Appellant has not renewed this argument before us.

[Footnote 10] In Fithian, the Supreme Court of California concluded that there was "no evidence that the application of California community property law interferes in any way with the administration or goals of the federal military retirement pay system. . . ." 10 Cal. 3d, at 604, 517 P.2d, at 457.

[Footnote 11] In Gorman, the California Court of Appeal held that Hisquierdo was based on the unique history and language of the Railroad Retirement Act of 1974; the court therefore considered itself bound to follow Fithian "pending further consideration of the issue by the California Supreme Court." 90 Cal. App. 3d, at 462, 153 Cal. Rptr., at 483. The California Supreme Court has since reaffirmed Fithian in In re Milhan, 27 Cal. 3d 765, 613 P.2d 812 (1980), cert. pending sub nom. Milhan v. Milhan, No. 80-578.

[Footnote 12] Appellee contends that this is not a proper appeal because appellant did not call the constitutionality of any statute into question in the California courts. Our review of the record, however, leads us to conclude otherwise. The Court of Appeal stated that appellant "also contends that the federal scheme of military retirement benefits pre-empts all state community property laws with respect thereto, and that California courts are accordingly precluded by the Supremacy Clause from dividing such benefits . . . ." App. to Juris. Statement 57. The court [453 U.S. 210, 220] flatly rejected this argument, id., at 57-59, and appellant then renewed it in his petition for hearing, p. 1, before the California Supreme Court. The present case thus closely resembles Dahnke-Walker Milling Co. v. Bondurant, 257 U.S. 282 (1921), where a state statute was challenged as being in conflict with the Commerce Clause. The Court held that the appeal was proper, since the appellant "did not simply claim a right or immunity under the Constitution of the United States, but distinctly insisted that as to the transaction in question the . . . statute was void, and therefore unenforceable, because in conflict with the commerce clause . . . ." Id., at 288-289. Accordingly, we conclude on the authority of Dahnke-Walker that this is a proper appeal. See also Japan Line, Ltd. v. County of Los Angeles, 441 U.S. 434, 440 -441 (1979).

[Footnote 13] In Tyler, the Court held that a retired officer was entitled to the benefit of a statute that increased the pay of "commissioned officers." The Court reasoned:

"It is impossible to hold that men who are by statute declared to be part of the army, who may wear its uniform, whose names shall be borne upon its register, who may be assigned by their superior officers to specified [453 U.S. 210, 222] duties by detail as other officers are, who are subject to the rules and articles of war, and may be tried, not by a jury, as other citizens are, but by a military court-martial, for any breach of those rules, and who may finally be dismissed on such trial from the service in disgrace, are still not in the military service." 105 U.S., at 246 (Emphasis in original.)

See also Kahn v. Anderson, 255 U.S. 1, 6 -7 (1921); Puglisi v. United States, 215 Ct. Cl. 86, 97, 564 F.2d 403, 410 (1977), cert. denied, 435 U.S. 968 (1978).

[Footnote 14] A retired officer may lose part of his retired pay if he takes Federal Civil Service employment. See 5 U.S.C. 5531 et seq. (1976 ed. and Supp. IV). He may lose all his pay if he gives up United States citizenship, see 58 Comp. Gen. 566, 568-569 (1979); accepts employment by a foreign government, U.S. Const., Art. I, 9, cl. 8, but see Pub. L. 95-105, 509, 91 Stat. 859 (granting congressional permission to engage in such employment with approval of the Secretary concerned and the Secretary of State); or sells supplies to an agency of the Department of Defense, or other designated agencies. 37 U.S.C. 801. See also Pub. L. 87-849, 2, 76 Stat. 1126 (retired officer may not represent any person in sale of anything to Government through department in whose service he holds retired status). The officer also may forfeit his retired pay if court-martialed. See Hooper v. United States, cited in the text.

[Footnote 15] Relying upon Tyler, the Ninth Circuit recently rejected the argument that Congress' alteration of the method by which retired pay is calculated deprived retired military personnel of property without due [453 U.S. 210, 223] process of law. Costello v. United States, 587 F.2d 424, 426 (1978), cert. denied, 442 U.S. 929 (1979). The court held that since "retirement pay does not differ from active duty pay in its character as pay for continuing military service," 587 F.2d, at 427, its method of calculation could be prospectively altered under the precedent of United States, v. Larionoff, 431 U.S. 864, 879 (1977). See also Abbott v. United States, 200 Ct. Cl. 384, cert. denied, 414 U.S. 1024 (1973); Lemly v. United States, 109 Ct. Cl. 760, 763, 75 F. Supp. 248, 249 (1948); Watson v. Watson, 424 F. Supp. 866 (EDNC 1976).

Some state courts also have concluded that military retired pay is not "property" within the meaning of their state divorce statutes because it does not have any "cash surrender value; loan value; redemption value; . . . [or] value realizable after death." Ellis v. Ellis, 191 Colo. 317, 319, 552 P.2d 506, 507 (1976). See Fenney v. Fenney, 259 Ark. 858, 537 S. W. 2d 367 (1976).

[Footnote 16] A number of state courts have held that military retired pay is deferred compensation, not current compensation for reduced services. See, e. g., In re Fithian, 10 Cal. 3d, at 604, 517 P.2d, at 456; In re Miller, ___ Mont. ___, 609 P.2d 1185 (1980), cert. pending sub nom. Miller v. Miller, No. 80-291; Kruger v. Kruger, 73 N. J. 464, 375 A. 2d 659 (1977). It is true that retired pay bears some of the features of deferred compensation. See W. Glasson, Federal Military Pensions in the United States 99 (1918). The amount of retired pay a service member receives is calculated not on the basis of the continuing duties he actually performs, but on the basis of years served on active duty and the rank obtained prior to retirement. See n. 7, supra. Furthermore, should the service member actually be recalled to duty, he receives additional compensation according to the active duty pay scale, and his rate of retired pay is also increased thereafter. 10 U.S.C. 1402, as amended by Pub. L. 96-342, 813 (b) (2), 94 Stat. 1102, and by Pub. L. 96-513, 511 (50), 94 Stat. 2924.

Nonetheless, the fact remains that the retired officer faces not only significant restrictions upon his activities, but also a real risk of recall. At [453 U.S. 210, 224] the least, then, the possibility that Congress intended military retired pay to be in part current compensation for those risks and restrictions suggests that States must tread with caution in this area, lest they disrupt the federal scheme. See Hooper v. United States, 164 Ct. Cl., at 159, 326 F.2d, at 987 ("the salary he received was not solely recompense for past services, but a means devised by Congress to assure his availability and preparedness in future contingencies"). Cf. Cong. Globe, 37th Cong., 1st Sess., 158 (1861) (remark of Sen. Grimes) (object of first non-disability retirement statute was "to retire gentlemen who have served the country faithfully and well for forty years, voluntarily if they see fit, (but subject, however, to be called into the service of the country at any moment that the President of the United States may ask for their services,) . . .").

[Footnote 17] Section 2771 provides in relevant part:

"(a) In the settlement of the accounts of a deceased member of the armed forces . . . an amount due from the armed force of which he was a member [453 U.S. 210, 225]   shall be paid to the person highest on the following list living on the date of death:

"(1) Beneficiary designated by him in writing to receive such an amount . . . .
"(2) Surviving spouse.
"(3) Children and their descendants, by representation.
"(4) Father and mother in equal parts or, if either is dead, the survivor.
"(5) Legal representative.
"(6) Person entitled under the law of the domicile of the deceased member."

Section 2771 was designed to "permit the soldier himself to

designate a beneficiary for his final pay." H. R. Rep. No. 833, 84th Cong., 1st Sess., 2 (1955). While this statute gives a service member the power of testamentary disposition over any amount owed by the Government, we do not decide today whether California may treat active duty pay as community property. Cf. Wissner v. Wissner, 338 U.S. 655, 657 , n. 2 (1950). We hold only that 2771, in combination with other features of the military retirement system, indicates that Congress intended retired pay to be a "personal entitlement."

[Footnote 18] An annuity under either plan is not "assignable or subject to execution, levy, attachment, garnishment, or other legal process." 10 U.S.C. 1440 and 1450 (i). Clearly, then, a spouse cannot claim an interest in an annuity not payable to him or her on the ground that it was purchased with community assets. See Wissner, 338 U.S., at 659 . Cf. Hisquierdo, 439 U.S., at 584 .

[Footnote 19] The RSFPP provides in relevant part:

"To provide an annuity under section 1434 of this title, a [service member] may elect to receive a reduced amount of the retired pay or retainer pay to which he may become entitled as a result of service in his armed force." 10 U.S.C. 1431 (b) (emphasis added).

The SBP states in relevant part:

"The Plan applies -
"(A) to a person who is eligible to participate in the Plan . . . and who is married or has a dependent child when he becomes entitled to [453 U.S. 210, 227]   retired or retainer

pay, unless he elects not to participate in the Plan before the first day for which he is eligible for that pay . . . ." 10 U.S.C. 1448 (a) (2) (1976 ed., Supp. IV) (emphasis added).

[Footnote 20] The SBP provides: "If a person who is married elects not to participate in the Plan at the maximum level or elects to provide an annuity for a dependent child but not for his spouse, that person's spouse shall be notified of the decision." 10 U.S.C. 1448 (a). But, as both the language of this section and the legislative history make clear, the spouse only receives notice; the decision is the service member's alone. See H. R. Rep. No. 92-481, at 8-9. An election not to participate in the SBP is in most cases irrevocable if not revoked before the date on which the service member first becomes entitled to retired pay. 1448 (a).

[Footnote 21] In Fithian, 10 Cal. 3d, at 600, 517 P.2d, at 454, the California Supreme Court observed and acknowledged: "Because federal military retirement pay carries with it no right of survivorship, the characterization of benefits as community property places the serviceman's ex-wife in a somewhat better position than that of his widow."

This is so for several reasons. If the service member does not elect to participate in the RSFPP or SBP, his widow will receive nothing. In contrast, if an ex-spouse has received an offsetting award of presently available community property to compensate her for her interest in the expected value of the retired pay, see n. 8, supra, she continues to be provided for even if the service member dies prematurely. See Hisquierdo, 439 U.S., at 588 -589. Furthermore, whereas an

SBP annuity payable to a surviving spouse terminates if he or she remarries prior to age 60, [453 U.S. 210, 228] see 10 U.S.C. 1450 (b), the ex-spouse's community awards against the retired service member continue despite remarriage. Lastly, annuity payments are subject to Social Security offsets, see 10 U.S.C. 1451, whereas community property awards are not. It is inconceivable that Congress intended these anomalous results. See Goldberg, Is Armed Services Retired Pay Really Community Property?, 48 Cal. Bar J. 89 (1973).

[Footnote 22] In addition, an Army enlisted man may not assign his pay. 37 U.S.C. 701 (c). While an Army officer may transfer or assign his pay account "[u]nder regulations prescribed by the Secretary of the [453 U.S. 210, 229] Army," he may do so only when the account is "due and payable." 701 (a). This limitation would appear to serve the same purpose as the prohibition against "anticipation" discussed in Hisquierdo, 439 U.S., at 588 -589. Cf. Smith v. Commanding Officer, Air Force Accounting and Finance Center, 555 F.2d 234, 235 (CA9 1977). But even if there were no explicit prohibition against "anticipation" here, it is clear that the injunction against attachment is not to be circumvented by the simple expedient of an offsetting award. See Hisquierdo, 439 U.S., at 588 . Cf. Free v. Bland, 369 U.S. 663, 669 (1962).

[Footnote 23] Appellee contends, mistakenly in our view, that the doctrine of non-attachability set forth in Buchanan simply "restate[s] the Government's sovereign immunity from burdensome garnishment suits . . . ." See Hisquierdo, 439 U.S., at 586 . Rather than resting on the grounds that garnishment would be administratively burdensome,

Buchanan pointed out: "The funds of the government are specifically appropriated to certain national objects, and if such appropriations may be diverted and defeated by state process or otherwise, the functions of the government may be suspended." 4 How., at 20. See also H. R. Rep. No. 92-481, at 17.

[Footnote 24] Under 814 of the Foreign Service Act of 1980, Pub. L. 96-465, 94 Stat. 2113, a former spouse who was married to a Foreign Service member for at least 10 years of creditable service is entitled to a pro rate share of up to 50% of the member's retirement benefits, unless otherwise provided by spousal agreement or court order; the former spouse also may claim a pro rate share of the survivor's annuity provided for the member's widow. Moreover, the member cannot elect not to provide a survivor's annuity without the consent of his spouse or former spouse.

The Committee Reports commented upon the radical nature of this legislation. See H. R. Rep. No. 96-992, pt. 1, pp. 70-71 (1980); S. Rep. No. 96-913, pp. 66-68 (1980); H. R. Conf. Rep. No. 96-1432, p. 116 (1980). During the floor debates Representatives Schroeder pointed out: "Whereas social security provides automatic benefits for spouses and former spouses, married at least 10 years, Federal retirement law has previously not recognized the contribution of the nonworking spouse or former spouse." 126 Cong. Rec. 28659 (1980). Representative Schroeder also noted that Congress had "thus far" failed to enact legislation that would extend to the military the "equitable treatment of spouses" afforded under the Civil Service and Foreign Service retirement systems. Id., at 28660.

[Footnote 25] Like the Foreign Service amendments, H. R. 2817, 96th Cong., 1st Sess. (1979), would have entitled a former spouse to a pro rata share of the retired pay and of the annuity provided to the surviving spouse; similarly, the bill would have required the service member to obtain the consent of his spouse and ex-spouse before electing not to provide a survivor's annuity. This bill was referred to the House Committee on Armed Services along with two other bills, H. R. 3677, 96th Cong., 1st Sess. (1979), and H. R. 6270, 96th Cong., 2d Sess. (1980). Whereas H. R. 2817 would have amended Title 10 to bring it into conformity with the Foreign Service model, these other two bills paralleled the Civil Service legislation, and would have authorized the United States to comply with the terms of a court decree or property settlement in connection with [453 U.S. 210, 232]   the divorce of a service member receiving retired pay. After extensive hearings, all three bills died in committee. See Hearing on H. R. 2817, H. R. 3677, and H. R. 6270 before the Military Compensation Subcommittee of the House Committee on Armed Services, 96th Cong., 2d Sess. (1980).

Legislation has been introduced in the 97th Congress that would require the pro rata division of military retired pay. See H. R. 3039, 97th Cong., 1st Sess. (1981), and S. 888, 97th Cong., 1st Sess. (1981). See also H. R. 3040, 97th Cong., 1st Sess. (1981) (pro rata division of retirement benefits of any federal employee).

[Footnote 26] A recent Presidential Commission has questioned the extent to which the military retirement system actually accomplishes these goals. See Report of the President's Commission on Military Compensation 49-56

(1978). Moreover, the Department of Defense has taken the position that service members are legally bound to comply with financial settlements ordered by state divorce courts; but while the Department did not oppose the legislation introduced in the 96th Congress that would have required the United States to honor community property divisions of military retired pay by state courts, it did express its concern over the dissimilar treatment afforded service members depending on whether or not they are stationed in community property States. See Hearing on H. R. 2817, H. R. 3677, and H. R. 6270 before the Military Compensation Subcommittee of the House Committee on Armed Services, 96th Cong., 2d Sess., 55, 58, 63 (1980) (statement of Deputy Assistant Secretary Tice). Of course, the questions whether the retirement system should be amended so as better to accomplish its personnel management goals, and whether those goals should be subordinated to the protection of the service member's ex-spouse, are policy issues for Congress to decide.

JUSTICE REHNQUIST, with whom JUSTICE BRENNAN and JUSTICE STEWART join, dissenting.

The Court's opinion is curious in at least two salient respects. For all its purported reliance on Hisquierdo v. Hisquierdo, 439 U.S. 572 (1979), the Court fails either to quote or cite the test for pre-emption which Hisquierdo established. In that case the Court began its analysis, after noting that States "lay on the guiding hand" in marriage law questions, by stating:

"On the rare occasion where state family law has come into conflict with the federal statute, this Court has limited review under the Supremacy Clause to a determination whether

Congress has `positively required by direct enactment' that state law be pre-empted. Wetmore v. Markoe, 196 U.S. 68, 77 (1904)." Id., at 581.

The reason for the omission of this seemingly critical sentence from the Court's opinion today is of course quite clear: the Court cannot, even to its satisfaction, plausibly maintain that Congress has "positively required by direct enactment" that California's community property law be pre-empted by the [453 U.S. 210, 237] provisions governing military retired pay. The most that the Court can advance are vague implications from tangentially related enactments or Congress' failure to act. The test announced in Hisquierdo established that this was not enough and so the critical language from that case must be swept under the rug.

The other curious aspect of the Court's opinion, related to the first, is the diverting analysis it provides of laws and legislative history having little if anything to do with the case at bar. The opinion, for example, analyzes at great length Congress' actions concerning the attachability of federal pay to enforce alimony and child support awards, ante, at 228-230. However interesting this subject might be, this case concerns community property rights, which are quite distinct from rights to alimony or child support, and there has in fact been no effort by appellee to attach appellant's retired pay. To take another example, we learn all about the provisions governing Foreign Service and Civil Service retirement pay, ante, at 230-232. Whatever may be said of these provisions, it cannot be said that they are "direct enactments" on the question whether military retired pay may be treated as community property. The conclusion is inescapable that the Court has no solid support for the conclusion it reaches -

What's Mine is Mine, What's Yours is Mine

certainly no support of the sort required by Hisquierdo - and accordingly I dissent.

Both family law and property law have been recognized as matters of peculiarly local concern and therefore governed by state and not federal law. In re Burrus, 136 U.S. 586, 593 -594 (1890); United States v. Yazell, 382 U.S. 341, 349 , 353 (1966). Questions concerning the appropriate disposition of property upon the dissolution of marriage, therefore, such as the question in this case, are particularly within the control of the States, and the authority of the States should not be displaced except pursuant to the clearest direction from Congress. [453 U.S. 210, 238]   Only in five previous cases has this Court found preemption of community property law. An examination of those cases clearly establishes that there is no precedent supporting admission of this case to the exclusive club.

The first such case was McCune v. Essig, 199 U.S. 382 (1905). McCune's father, a homesteader, died before completing the necessary conditions to obtain title to the land. McCune claimed that under the community property laws of the State of Washington she was entitled to a half interest in her father's land. Congress in the Homestead Act, a however, had "positively required by direct enactment," Hisquierdo, supra, at 581, that in the case of a homesteader's death the widow would succeed to the homesteader's interest in the land. Indeed, the Act set forth an explicit schedule of succession which specifically provided for a homesteader's daughter such as McCune. She succeeded to rights and fee under the statute only in the case of the death of both her father and mother. In the words of Justice McKenna:

"It requires an exercise of ingenuity to establish uncertainty in these provisions. . . . The words of the statute are clear, and express who in turn shall be its beneficiaries. The contention of appellant reverses the order of the statute and gives the children an interest paramount to that of the widow through the laws of the state." 199 U.S., at 389.

There is, of course, nothing remotely approaching this situation in the case at bar. Congress has not enacted a schedule governing rights of ex-spouses to military retired pay and appellee's claim does not go against any such schedule. 1 [453 U.S. 210, 239]

The next case from this Court finding pre-emption of community property law did not arise until 45 years later. In Wissner v. Wissner, 338 U.S. 655 (1950), the deceased serviceman's estranged wife claimed she was entitled to one-half of the proceeds of a National Service Life Insurance policy, the premiums of which were paid out of the serviceman's pay accrued while he was married, even though decedent had designated his parents as the beneficiaries. The Act in question specifically provided that the serviceman shall have "'the right to designate the beneficiary or beneficiaries of the insurance [within a designated class]. . . . and shall . . . at all times have the right to change the beneficiary or beneficiaries.'" Id., at 658 (quoting 38 U.S.C. 802 (g) (1946 ed.)). As the Court interpreted this. "Congress has spoken with force and clarity in directing that the proceeds belonged to the named beneficiary and no other." 338 U.S., at 658 . That is not at all the case here. Congress has provided that the serviceman receive retired pay in 10 U.S.C. 3929, to be sure, but that is simply the general provision permitting payment - it hardly evinces the "deliberate purpose of Congress" concerning the question before us, as was the case with the designation of a life insurance policy beneficiary in Wissner. 338 U.S., at 659 .

The Court in Wissner also noted that the statute provided that "[p]ayments to the named beneficiary `shall be exempt [453 U.S. 210, 240] from the claims of creditors, and shall not be liable to attachment, levy, or seizure by or under any legal or equitable process whatever, either before or after receipt by the beneficiary.'" Ibid. (quoting 38 U.S.C. 816 (1946 ed.)). The wife's claim was thus in "flat conflict" with the terms of the statute. 338 U.S., at 659 . This forceful and unambiguous language protecting the rights of the designated beneficiary has no parallel so far as military retired pay is concerned.

It is important to recognize that the Court's analysis, while purporting to rely on Wissner, actually is contrary to the analysis in that case. As will be explored in greater detail below, the Court focuses on two provisions in concluding that military retired pay cannot be treated as community property: the provision permitting a serviceman to designate who shall receive any arrearages in pay after his death, and the provision permitting a retired serviceman to fund an annuity for someone other than the ex-spouse out of retired pay. The Court's theory is that since the serviceman can dispose of part of the retired pay without participation of the ex-spouse - either the arrearages or the premiums to fund the annuity - the retired pay cannot be treated as community property. This, however, is precisely the analysis the Wissner court declined to adopt in concluding that the proceeds of an insurance policy, purchased with military pay, could not be treated as community property. The Wissner court simply concluded that the wife could not pursue her community property claim to the proceeds, even though purchased with community property funds. This is comparable to ruling in this case that appellee cannot obtain half of any annuity funded out of retired pay pursuant to the statute, or half of the arrearages, when the serviceman has designated someone else to receive them. The Wissner court specifically left open the question whether the whole from which the premiums were taken - the military pay - could be treated as community property. Id., at 657. n. 2. That is, however, the analytic jump the Court takes today, in ruling that retired pay cannot [453 U.S. 210, 241] be treated as community property simply because parts of it, or proceeds of parts of it - arrearages and the annuity - cannot be. 2

The next two cases, Free v. Bland, 369 U.S. 663 (1962), and Yiatchos v. Yiatchos, 376 U.S. 306 (1964), involved the same provisions. Plaintiffs sought community property rights in United States Savings Bonds, even though duly issued Treasury Regulations provided that designated co-owners would, upon the death of the other co-owner, be "the sole and absolute owner" of the bonds. No such language is involved in this case.

The most recent case is, of course, Hisquierdo, in which the Court held that Congress in the Railroad Retirement Act preempted community property laws so that a railroad worker's pension could not be treated as community property. It bears noting that this case is not Hisquierdo revisited. In Hisquierdo there was a specific statutory provision which satisfied the requirement that Congress "'positively requir[e] by direct enactment' that state law be pre-empted." 439 U.S., at 581 (quoting Wetmore v. Markoe,

196 U.S. 68, 77 (1904)). Section 14 of the Railroad Retirement Act of 1974, carrying forward the provisions of 12 of the Act of 1937, provided:

"Notwithstanding any other law of the United States, or of any State, territory, or the District of Columbia, no annuity or supplemental annuity shall be assignable or be subject to any tax or to garnishment, attachment, or other legal process under any circumstances whatsoever, nor shall the payment thereof be anticipated." 45 U.S.C. 231m. [453 U.S. 210, 242]

The Hisquierdo Court viewed this provision as playing "a most important role in the statutory scheme." 439 U.S., at 583 -584. The Court stressed the language "[n]otwithstanding any other law . . . of any State." id., at 584, and noted that 14 "pre-empts all state law that stands in its way." Ibid.

With all the emphasis placed on 14 in Hisquierdo, one would have expected the counterpart in the military retired pay scheme to figure prominently in the Court's opinion today. There is, however, nothing approaching 14 in the military retired pay scheme. The closest analogue, 37 U.S.C. 701 (a), is buried in footnote 22 of the Court's opinion. It simply provides:

"Under regulations prescribed by the Secretary of the Army or the Secretary of the Air Force, as the case may be, a commissioned officer of the Army or the Air Force may transfer or assign his pay account, when due and payable."

The contrast with the provision in Hisquierdo is stark.

187

Section 14 forbids assignment; 701 (a) permits it. Section 14 contains a "flat prohibition against attachment and anticipation," 439 U.S., at 582 ; all that can be gleaned from 701 (a) is a negative implication prohibiting voluntary assignments prior to the time pay is due and payable. Such a limit is of course a far cry from the Hisquierdo provision requiring that the retired pay may not be subject to "legal process under any circumstances whatsoever" and that it shall not "be anticipated." It is no wonder 701 (a) is buried in a footnote in the Court's opinion. 3 [453 U.S. 210, 243]

In addition to 14 the Hisquierdo Court also relied on the fact that the Railroad Retirement Act provided a separate spousal entitlement. "embod[ying] a community concept to an extent." 439 U.S., at 584 . Under the Railroad Retirement Act, 45 U.S.C. 231d (c), a spouse is entitled to a separate benefit, which terminates upon divorce. 231d (c) (3). Congress explicitly considered extending the spousal benefit to a divorced spouse but declined to do so. 439 U.S., at 585 . The Hisquierdo Court found support in this not to permit California to expand the community property concept beyond its limited use by Congress in the Act. No similar separate spousal entitlement, terminable on divorce, exists in the statutes governing military retired pay. The "this far and no further" implication in Hisquierdo, therefore, cannot be made here.

II

The foregoing demonstrates that today's decision is not simply a logical extension of prior precedent. That does not, to be sure, mean that it is necessarily wrong - there has to be a first time for everything. But examination of the analysis in the Court's opinion convinces me that it is both unprecedented and wrong.

In its analysis the Court contrasts the statute involved in Hisquierdo, noting that there spouses received an annuity which terminated upon divorce. Here there is no such provision. As the Court states its conclusion: "Thus, unlike the Railroad Retirement Act, the military retirement system does not embody even a limited `community property concept.'" Ante, at 224. This analysis, however, is the exact opposite [453 U.S. 210, 244] of the analysis employed in Hisquierdo. As we have seen, there the Court's point was that Congress had provided some community property rights and made a conscious decision to provide no more:

"Congress carefully targeted the benefits created by the Railroad Retirement Act. It even embodied a community concept to an extent . . . . Congress purposefully abandoned that theory, however, in allocating benefits upon absolute divorce. . . . The choice was deliberate." 439 U.S., at 584 - 585.

Now we are told that pre-emption of community property law is suggested in this case because there is no community property concept at all in the statutory scheme. Under Hisquierdo, this absence would have been thought to suggest that there was no pre-emption, since the argument could not be made, as it was in Hisquierdo, that Congress had addressed the question and drawn the line. See In re Milhan, 27 Cal. 3d 765, 775-776, 613 P.2d 812, 817 (1980), cert. pending sub nom. Milhan, v. Milhan, No. 80-578. I am not certain whether the analysis was wrong in Hisquierdo or in this case, but it is clear that both cannot be correct. One is led to inquire where this moving target will next appear.

The Court also relies on "several features of the statutory scheme" as evidence that Congress intended military retired pay to be the "personal entitlement" of the serviceman. The Court first focuses on 10 U.S.C. 2771, which permits a serviceman to select the beneficiary of unpaid arrearages. As we have seen, supra, at 240-241, the Court's reliance on Wissner in this context establishes, at most, only that unpaid arrearages cannot be treated as community property, not that retired pay in general cannot be. A provision permitting a serviceman to tell the Government where to mail his last paycheck after his death hardly supports the inference of a congressional intent to pre-empt state law governing disposition of military retired pay in general. [453 U.S. 210, 245]

The Court next relies on the statutory provisions permitting a retired serviceman to fund an annuity for his potential widow and/or dependent children out of retired pay. Even granting the Court its premise that the annuity is not subject to community property treatment, the conclusion that military retired pay is not subject to community property treatment simply does not follow. If California's community property law conflicts with permitting a retired serviceman to fund an annuity out of retired pay, then by all means override California's law - to the extent of the conflict. Even if Congress did intentionally intrude on community property law to the extent of permitting a serviceman to fund an annuity, that hardly supports an intent to intrude on all community property law. Nothing in the Court's analysis shows any reason why appellee should not be entitled to one-half of appellant's retired pay less amounts he uses to fund an annuity, should he decide to do so.

The Court resists the recognition of any rights to retired pay in the ex-spouse because of a policy judgment that it would be "anomalous" to place the ex-spouse in a better position than a widow receiving benefits under an annuity. Ante, at 227. The Court, however, is comparing apples and oranges in two respects. The ex-spouse's rights are to retired pay, and cease when the serviceman dies. The widow's rights are to an annuity which begins when the serviceman dies. The fact that Congress "deliberately has chosen to favor the widower or widow over the ex-spouse" so far as the annuity is concerned, ante, at 228, simply has no relevance to the rights of the ex-spouse to the retired pay itself. Second, the ex-spouse has contributed to the earning of the retired pay to the same degree as the serviceman, according to state law. The widow may have done nothing at all to "earn" her annuity, as would be the case, for example, if appellant remarried and funded an annuity for his widow out of retired pay. In view of this, I see nothing "anomalous" in providing the ex-spouse with rights in retired pay. In any event, such policy [453 U.S. 210, 246] questions are for Congress to decide, not the Court, and the Court fails in its efforts to show Congress has found California's system anomalous.

The third argument advanced by the Court is the weakest of all: the Court argues that an ex-spouse in a community property State cannot obtain half of the military retired pay, by attachment or otherwise, because she can obtain alimony and child support by attachment. This is pre-emption by negative implication - not the "positive requirement" and "direct enactment" which Hisquierdo indicated were required. And since appellee does not seek to attach anything, even the negative implication is not directly relevant.

The Court also stresses the recognition of community property rights in varying degrees in the Foreign Service and Civil Service laws. Again, this hardly meets the Hisquierdo test. Both the Foreign Service and Civil Service laws are quite different from the military retired pay laws. The former contain strong anti-attachment provisions like 14 of the Railroad Retirement Act considered in Hisquierdo, see 5 U.S.C. 8346; 22 U.S.C. 1104, so Congress could well have thought explicit legislation was necessary in these areas.

III

The very most that the Court establishes, therefore, is that the provisions governing arrearages and annuities pre-empt California's community property law. There is no support for the leap from this narrow pre-emption to the conclusion that the community property laws are pre-empted so far as military retired pay in general is concerned. Such a jump is wholly inconsistent with . this Court's previous pronouncements concerning a State's power to determine laws concerning marriage and property in the absence of Congress' "direct enactment" to the contrary, and I therefore dissent.

[Footnote 1] The Court maintains that the present case is like McCune: "[s]o here, the right appellee asserts 'reverses the order of the statute' by giving the ex-spouse an interest paramount to that of the surviving spouse and children of the service member . . . ." Ante, at 233. With all respect, I do not understand the statute to establish any ordered list of those with interests in retired pay. The Court's argument is apparently that [453 U.S. 210, 239] recognizing the ex-spouse's interest in retired pay would burden the serviceman's decision to fund an annuity for his current spouse out of retired pay. This is of course a far cry from the situation in McCune, where the statute accorded the surviving widow and daughter specific places and the daughter sought to switch the order by invoking community property law. Even if the Court is correct that there is a conflict between California's community property law and the decision of the serviceman to fund an annuity out of retired pay, the answer is not to pre-empt community property treatment across the board, but only to the extent of the conflict, i. e., permit community property treatment of retired pay less any amounts which are used to fund an annuity. See infra, at 245.

[Footnote 2] The error in the Court's logic is perhaps most apparent when it is recognized that the arrearages provision applies to regular military pay as well as retired pay. The Court's logic would compel the conclusion that regular pay is thus not subject to community property treatment, an untenable position which the Court itself shies away from without explanation, ante, at 224-225, n. 17.

[Footnote 3] The Court states that "[r]etired pay cannot be attached to satisfy a property settlement incident to the dissolution of a marriage," ante, at 228. The sources for this are not statutory but rather a common-law doctrine, Buchanan v. Alexander, 4 How. 20 (1845), and a House Report explaining a decision not to enact a bill, see ante, at 228-230. The Court cannot of course justify either source as Congress "positively requir[ing] by direct [453 U.S. 210,

243]   enactment" that state law be pre-empted. See Hisquierdo, 439 U.S., at 581 . Thus even accepting the rule, it does not, as 14 of the Railroad Retirement Act did in Hisquierdo, evince the strong congressional intent that military retired pay "actually reach the beneficiary." And congressional intent is all the prohibition on attachment is relevant to, since appellee seeks neither anticipation of pay nor attachment from the Government. [453 U.S. 210, 247]

For more information and the most recent developments visit the ULSG website at:

www.ULSG.org